Salt Affected Soils:
Reclamation and Management

Related Books

Title	Author
• A Manual on Conservation of Soil and Water	*USDA*
• Agriculture and Fertilizers	*O.C. Bockman*
• Agrochemicals in Plant Disease Management	*N.G. Ravichandran*
• Agrotechnology for Dryland Farmaing *2nd Revised Ed*	*A.M. Dhopte*
• Biofumigation and Solarization for Management of Soil-Borne Plant Pathogens	*P.P. Reddy*
• Crop Production in Waterlogged Saline Soils	*S.K. Gupta*
• Crop Production in Salt Affected Soils	*I.C. Gupta*
• Crop Protection Strategies under Climate Change Scenarios	*P.P. Reddy*
• Drainage Engineering: Principles and Practices	*S.K. Gupta*
• Fertilizer Manual	*IFDC*
• Fundamentals of Organic Farming and Gardening: An Instructors Guide	*A. McGregor*
• Fundamentals of Soil Science	*A. Rathinasamy*
• Glimpse on General Agriculture	*P. Laxman Rao*
• Handbook of Saline and Alkali Soils Diagnosis Reclamation and Management	*S.K. Gupta*
• Manual on Fundamentals of Agronomy	*L.K. Jain*
• Nanotechnology	*J.C. Tarafdar*
• Organic Farming Theory & Practice	*S.P. Palaniappan*
• Principles in the Quantitative Analysis of Water Fertilizers Plants and Soils	*U.S.S. Ramulu*
• Rainwater Management Theory and Practice	*M.L. Jat*
• Salinity Tolerance in Plants: Methods, Mechanisms and Management *2nd Ed*	*B.K. Garg*
• Salt Affected Soils: Reclamation and Management	*S.K. Gupta*
• Soil and Plant Analysis	*C.S. Piper*
• Soil and Water Conservation in Semi-Arid Areas	*N.W. Hudson*
• Soil Chemical Analysis	*M.L. Jackson*
• Soil Conservation Technical Handbook	*D.H. Hicks*
• Soil Guide: A Handbook for Understanding and Managing Agricultural Soils	*Geoff Moore*
• Studies in Arid Land Management	*T.S. Chouhan*
• Sustainable Agriculture	*R.S. Meena*
• Sustainable Development of Dryland Agriculture in India	*R.P. Singh*
• The Use of Saline Waters for Crop Production	*FAO*
• Water Measurement Manual *3rd Ed*	*USDI*

Readers use promo code: **SPNEW10** to get offer on above books

Salt Affected Soils: Reclamation and Management

S. K. Gupta
I. C. Gupta

SCIENTIFIC PUBLISHERS

Published by

SCIENTIFIC PUBLISHERS (INDIA)
5 A, New Pali Road, P.O. Box 91
Jodhpur 342 001 (INDIA)
E-mail: info@scientificpub.com
Website : www.scientificpub.com

Print : 2019

ISBN: 978-81-72338-60-2
ISBN: 978-93-86237-51-4

Printed in India

PREFACE

To ensure equitable distribution of water and to meet the ever increasing demand for domestic, industrial and agricultural sectors of the arid and semiarid region, India decided to develop its water resources in an integrated manner. The approach called for storage, diversion and transportation of water from excess to deficit water regions. The approach initially led to increased production and productivity and provided needed food security to the nation. The benefits reaped as a result of this approach are not unblemished. The most conspicuous adverse environmental effects appeared in the form of water logging and soil salinity adversely impacting the productivity of the land and the water resources. The vast stretches of once productive agricultural lands went out of cultivation leading to the development of 'wet deserts'.

The problem of water logging and soil salinity is not confined to irrigated lands but large chunks of land in coastal sub-humid and humid regions are also affected by twin problems due to varied reasons. It has been estimated that overall about 6.73 million ha of agricultural land is suffering due to water logging and/or soil salinity. Researchers in the country have come out with sound technical and economically viable technologies for the reclamation and management of waterlogged salt affected lands. These technologies have been well received by the stakeholders and the planners. More than 1.85 million ha of waterlogged salt affected soils have so far been reclaimed. The food grains produced on these lands is valued at Rs. 13.5 billion per annum. However, the technical breakthrough achieved in the reclamation and management of waterlogged saline soils has not yet been translated on a desirable scale more so in far flung areas habited by resource poor communities. It could be due to non-availability of these technologies in practical terms or due to resource crunch.

Realizing this, an attempt was made by the authors to collate the existing technologies and put them in a print form so that people could grasp the intricacies of these technologies. The book titled "Crop Production in Waterlogged Saline Lands" was published more than 10 years hence. Since the publication of this book, many innovative cost effective solutions have emerged to reclaim waterlogged salt affected soils. In order to provide the latest information it has been decided to thoroughly revise the book and spread its canvass to include reclamation of both the saline and the alkali soils. All the chapters have been fully revised and reorganized to address the new challenges and their solutions. The title of the book has also been revised to Salt Affected Soils : Reclamation and Management.

The first chapter of the book discusses the extent, distribution and nature of the problems including diagnostic analysis procedures. Technologies to improve field drainage in humid/sub-humid and semiarid/arid regions are discussed in chapter 2. The basics of soil salinization of the root zone, which is often cited as the root cause of environmental degradation in irrigation commands forms the subject matter of chapter 3. The principles of leaching of saline/alkali or saline water irrigated lands also form the part of this chapter. Principles and package of practices to reclaim alkali, inland saline and coastal saline soils have been included in chapters 4, 5 and 6 respectively. The Differences in approaches to reclaim these kinds of soils have been explicitly highlighted. Finally the socio-economic aspects to decide the economic viability and social acceptability of reclamation projects have been included in chapter 7.

We hope that this book will be immensely useful to researchers, teachers and students, officers of command area development authorities, extension workers and above all the farming community of India. Any progress made in land reclamation programs with the help of this book will give us immense satisfaction. We would welcome suggestions to improve upon the contents and layout of the book.

S.K. Gupta
I.C. Gupta

CONTENTS

1

EXTENT, DISTRIBUTION AND NATURE OF PROBLEM SOILS

SOIL RESOURCES OF INDIA

The land mass, which forms the subcontinent of India, covers an area of 329 million ha (M ha). It constitutes about 2.3% of the global land mass, as against the 17% of the world's population. Soil survey reports of India have identified more than 26 soil types. Rayachaudhari (1962) quantitatively estimated the area under different soil groups (Table 1.1) and observed that the alluvial soils, the black soils, the red soils and the laterite and lateritic soils are the major soil groups of significance.

Alluvial soils is by far the largest and the most important soil group spread over the Indo-Genetics plains, the Brahmaputra valley and deltas of several major rivers. Newer alluviums are sandy and have less calcium carbonate content and are lighter in color. Older alluviums are more clayey in composition and have a layer of $CaCO_3$ concretion at shallow depth (1.0-1.5 m). Nitrogen deficiency is exhibited to a high degree in most of the alluvial soils and is a major limiting factor in crop production. High exchangeable sodium and/or high salt content also limit crop production in saline alkali soils.

Black soils are the typical soils derived from the Deccan trap as a result of weathering of a variety of parent materials. Deep black soils derived from trap basalt, gneiss, shales, schists and limestone are often of mixed origin. Shallow black soils are lighter in color and have lime layer exposed to the surface. The black soils are fairly fine textured with clay content between 35 and 60%. These soils contain moderate amount of salts. It is likely that these soils may turn saline/alkali upon introduction of irrigation.

Table 1.1 Area under different soil types

Type of soils	Approximate area (M ha)
Alluvial soils	101.2
Alluvial soils highly calcareous	8.9
Coastal alluvium	8.5
Deltaic alluvium (occasionally saline)	17.0
Alluvial soils affected by salinity and alkalinity	6.9
Desert soils	14.6
Deep black soils	6.9
Medium black soils	18.6
Shallow black soils	4.9
Black soils affected by salinity and alkalinity	6.9
Black soils undifferentiated	12.5
Mixed red and black soils	10.5
Red soils	30.4
Red gravelly soils	1.6
Red and yellow soils	17.8
Laterite soils	10.1
Laterite and lateritic soils	2.0
Brown soils under deciduous	1.6
Grey and brown soils	3.6
Hill soils	2.4
Podsolic soils	3.6
Forest soil laterised	6.5
Foothill *tarai* hill soils	5.7
Mountain meadow soils	11.7
Skeletal soils	2.4
Peat muck, very humus and humus soils also called bog soils, organic soils and half bog soils	0.2
Other soils (viz. mountain soils undifferentiated glaciers and eternal snow)	11.8
Total	**328.8**

The ancient crystalline and metamorphic rocks on meteoric weathering have given rise to what is called red soil. These soils are derived from granite, gneiss sandstone, basalt and shales and have originated from weathering of crystalline and metamorphic and other sub-ordinate rocks. These are rich in ferro magnesium minerals and have kaolinite predominance. A diagnostic argillic horizon develops in these soils as a result of clay transport. Depending upon the vegetation and topography, red soils differ in texture and color.

Laterite and lateritic soils are the formations peculiar to India and some other tropical countries. It is a compact to vesicular rock composed essentially of a mixture of the hydrated oxides of aluminum and ions with small amounts of manganese oxide. All laterite soils are poor in lime and magnesia and deficient in nitrogen. Occasionally the phosphate content may be high.

Many other groups of soils also occupy extensive areas in the country. More details about these soils can be found in Soils of India and their Management (FAI, 1985) and Benchmark Soils of India (Murthy, 1982).

Problems Associated with Land Resource

Land is the basic resource base of production in agriculture, forestry, animal husbandry etc. While demand for food, fiber and fodder is increasing as a result of growing human and livestock population, the vital soil resource is almost inelastic. The net sown area of about 119 M ha in 1950 expanded to 143 M ha in 1980 and is likely to go up to 155 M ha in 2025. With increasing demand from non agricultural sectors, such as residential, industrial, recreational and areas needed for development activities such as roads, railways and irrigation/drainage systems, it is most likely that area for horizontal expansion might not be available and the area may continue to hover around 140-145 M ha. Thus, any increase in agricultural production would be governed by our efforts to manage the existing soil and water resources to obtain maximum production per unit of these resources.

Even the currently available land resource of the country is afflicted with a number of problems causing overall poor yields. According to Department of Agriculture and Cooperation, out of a total geographical area of 328.74 M ha, 173.64 M ha is being degraded by one or the other factor. Using Global Assessment of Soil Degradation guidclines, the NBSS&LUP, Nagpur has assessed that 187.8 M ha has been degraded by various degradation processes to various degrees (Table 1.2).

Table 1.2 India's land degradation statistics

Kind of degradation	Area affected (M ha)	% of total
Water erosion	149.037	45.3
Wind erosion	13.489	4.1
Chemical deterioration	13.818	4.2
Physical deterioration	11.515	3.5
Not fit for agriculture	18.095	5.5
Unaffected area	123.046	37.4

Source: Sehgal and Abrol (1994)

WATER RESOURCES OF INDIA

The country as a whole receives an average annual rainfall of about 120 cm which over the total land area of 329 M ha yields an annual precipitation of 400 M ha-m including some net inflow from neighboring countries. A percentage of this rainfall goes as runoff in the river systems; part is retained in the soil while some recharges the ground water aquifer.

Surface Water Resources

To estimate the surface water resources of India, the country has been divided into six river basins, namely Indus basin, Ganga basin, Brahmaputra basin, East Coast, West Coast and the Rajputana region. The First Irrigation Commission estimated surface water resources of the country as 144.32 M ha-m. Later on it was revised to 167.23 M ha-m. Statistical estimate of the Central Water and Power Commission put this figure at 188.21 M ha-m. A comprehensive assessment made by the National Commission on Agriculture (1976) indicated that surface water resources of the country are 178 M ha-m. According to the present indications, the total and utilizable surface water resources of the country are 176.8 M ha-m and 70 M ha-m respectively. An upward revision in the utilizable flows may be expected on account of advancement in technology and on account of large scale inter-basin transfer of water.

Ground Water Resources

Raghava Rao *et al.* (1969) estimated that about 42.6 M ha- m of replenishable ground water resources are available in the country. This estimate was also endorsed by the Task Force on Ground Water

Resources of the Planning Commission. Considering losses from evapotranspiration, subsurface flows, recharge in the saline water regions and efficiency etc., the Task Force recommended that on a long-term basis, a potential of 20.36 M ha-m of replenishable ground water exists for exploitation. This does not include ground water in storage. According to the Central Ground Water Board (Government of India), the country has an exploitable ground water potential of 32.3 M ha-m.

A major limitation for the use of ground water has been its quality. As innumerable factors affect the ground water quality, wide variations in water quality are observed within the country and within a state. The ground water quality surveys have indicated that the poor quality waters constitute a major portion (25- 84%) of the total irrigation potential of ground water in arid and semiarid regions (Table 1.3).

Table 1.3 Ground water potential and percent distribution of marginal and poor quality ground water in arid and semiarid regions of India

State	Utilizable potential (M ha-m/yr)	Net draft (M ha-m/ annum)	Distribution of ground water (%)		
			Good	Marginal	poor
Andhra Pradesh	3.66	0.74	68	25	7
Gujarat	2.03	0.69	70	20	10
Haryana	0.88	0.61	37	8	55
Karnatka	1.30	0.18	62	12	26
Madhya Pradesh	5.95	0.49	75	10	15
Punjab	1.31	0.93	59	22	19
Rajasthan	1.83	0.43	16	16	68
Tamil Nadu	2.69	0.99	*	*	*
Uttar Pradesh	9.27	2.68	32	22	46
Total	**28.92**	**7.74**	**419**	**135**	**246**

* Not estimated

Utilizable Flows and Irrigation Potential

The utilizable flows aggregate to about 105 M ha-m out of which 70 M ha-m are from surface and 35 M ha-m from the ground water resource. It may be understood that not all the water will be available for irrigation as there are other demands on it. It is expected that in 2025

AD, about 77 M ha-m of water would be available for irrigation and the rest would be for domestic and industrial uses. The net sown area in 2025 AD is estimated to be 155 M ha. With anticipated cropping intensity of 136%, the gross cropped area would be about 210 M ha. The available water resources even at that time would only be sufficient to provide irrigation to an area of about 110 M ha. Therefore, with limited resources, it is expected that only 52% of the gross sown area of 210 M ha will be under irrigation in 2025. The remaining area will continue to depend upon the rainfall.

Table 1.4 Irrigation potential created and utilized (M ha)

Plan	**Potential created**			**Potential utilized**
	Maj. & Med.	**Minor**	**Total**	**Total**
At the end of Pre Plan up to 1951	9.7	12.9	22.6	22.6
First (1951- 56)	12.2	14.0	26.2	25.0
Second (1956-61)	14.3	14.7	29.0	27.8
Third (1961-66)	16.6	17.0	33.6	32.2
Annual Plans (1966-69)	18.1	19.0	37.1	35.8
Fourth (1969-74)	20.7	23.4	44.1	41.9
Fifth (1974-78)	24.7	27.3	52.0	48.5
Annual Plans (1978-80)	26.6	30.0	56.6	52.6
Sixth (1980-85)	27.7*	37.5	65.2	58.8
Seventh (1985-90)	29.9	46.6	76.5	68.6
At the end of 1991-92	30.7	50.3	81.0	72.9
Eighth (1992-97)	33.0	53.3	86.3	77.2
Ninth (1997-2002)	37.1	56.9	94.0	81.0
Tenth (2002-07)	42.4	60.4	102.8	87.2
Eleventh (2007-12)	-	-	108.2	

* Reappraised figures

To usher in an era of self sufficiency in food production, expansion of irrigation was thought to be essential and in pursuance of this view, 246 major, 1039 medium and 1.08 million minor irrigation projects to harness surface water were taken up between 1951 and 1990. With more than 17 million wells nationwide, ground water now supplies water to more than 50% of the irrigated area and due to higher yields in ground water irrigated areas, even higher proportion of the total irrigated output. According to some estimates, 70-80% of the value of irrigated production in India may depend on ground water irrigation. Irrigation potential created increased from 22.6 M ha in 1951 to 108.21 M ha by March 2010, i.e. at the end of the first three years of 11th plan (Table 1.4; Publication Division (2011), CWC (2010)). However, this achievement was not unblemished. In the wake of increased productivity and improved quality of life, there were some adverse environmental impacts in the form of water logging and soil salinization of agricultural lands. These are most serious environmental problems faced by the farmers in the irrigation commands of India.

IRRIGATION AND WATER LOGGING

Irrigation is an art which has been practiced in India since time immemorial. In Narada Smrity, it is mentioned that *Rishi* Narada inquired from emperor Yudhisthra whether the farmers were sturdy and prosperous and whether the dams had water for distribution in different parts of the kingdom. The remains of Indus Valley Civilization and Vedic texts bear testimony to the practice of irrigation. The Grand *Anicut* across river Cauvery in Tamil Nadu was constructed in the second century A.D (Fig. 1.1 left). Feroze Shah Tuglaq in fourteenth century AD got the Western Yamuna Canal constructed (Fig. 1.1 right). Irrigation works such as construction of a large number of wells all over the country, tanks in south India and inundation canals in north India were constructed during early nineteenth century. The total irrigated area through public works in 1900 stood at 13.3 M ha which increased to 19.3 M ha in 1920-21. At the time of partition, the total irrigated area stood at 28.2 M ha out of which 8.8 M ha went to Pakistan leaving 19.4 M ha with India. Before the plan period started in 1950, the irrigated area had already increased to 22.6 M ha. The irrigated area rose to 102.8 M ha in 2006-07. It is a matter of concern that the irrigation potential created has not been fully utilized. The gap between development and utilization has been estimated to be 15.6 M ha at the end of 10th Plan (Table 1.4).

Inter basin transfer of water to irrigate lands in arid and semiarid regions interferes with the hydrological cycle. It creates imbalance between the recharge to the ground water and the discharge through the natural out flow. As a result, water is taken into storage leading to rise in the water table. It is, therefore, believed that introduction of irrigation and development of water logging and soil salinity is closely interlinked. "Turning white of fields" has been mentioned in a Babylonian story of

Atrasis (1800 BC). In the Indian context, harmful effects of excess water are enunciated in the epic Narada Smrity where sage Narada says, "No grain is ever produced without water, but too much water tends to spoil the grain". An inundation is as injurious to growth as dearth of water. Practically, however, there is no mention of the problem till 19th century when the problems of water logging and soil salinity attracted the attention of the Government.

Fig. 1.1 Grand *Anicut* across river Cauvery (left) and Western Yamuna canal (right)

Western Yamuna canal, constructed in 14th century, was realigned by the British and opened for regular irrigation in 1839. The complaints of soil salinization first appeared in a village Munak in Punjab (Now in Haryana) in the year 1855; 16 years after the system became operational. It was not an isolated case. In the year 1876, reports of water logging and soil salinity emerged from Uttar Pradesh when an indigo farmer asked for compensation in lieu of the damage to his land on account of irrigation. Irrigation was introduced in the Nira canal command in Maharashtra during 1885. Reports of water logging and soil salinity appeared in the year 1888 and by 1905 an area of 1,600 ha out of an irrigated area of 14,800 ha was taken out of cultivation. The problem areas increased to 5,570 ha in 1947. Twin problems of water logging and soil salinity were also observed within few years after the opening of the lower Chenab canal in 1892.

Recent experiences on this score also reveal an inter link between the introduction of irrigation and spread of water logging and soil salinity. Projects, commissioned in the post independence period, also developed serious water logging and soil salinity problems. The water logging developed so fast in Tawa project in Madhya Pradesh that within few years of opening of the canal, farmers formed "*Mitti Bachao Abhiyan Samities*" (Save the Soil Committees). The water table in an area of about 100,000 ha rose at an alarming rate in the Phase I of the Indira Gandhi *Nahar Pariyojana* (Fig. 1.2). Even today the picture is not very rosy on that score. The water table at Bhuna in Haryana in the Bhakra Canal Command rose @ 0.6 m per annum during 1975-2000. The problems of water logging and soil salinity increased @ 3000 ha per annum in the Tungabhadra canal command in Karnataka during 1979-1997 (Fig. 1.3). Even in the medium lift irrigation project undertaken by the farmer's societies, water logging and soil salinity has appeared in the state of Maharashtra. This calls for introduction of preventive technology at the time of planning of an irrigation project, close monitoring of the project and implementing drainage projects as soon as the problem shows first sign of appearance. Some more examples of such a scenario appear in a forthcoming sections of this chapter.

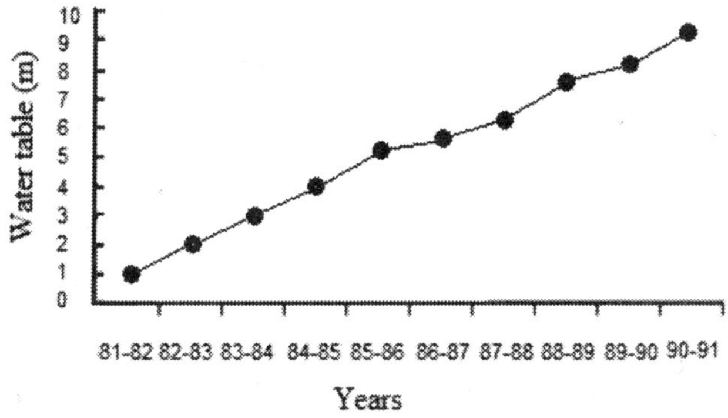

Fig. 1.2 Trend of water table rise in Indira Gandhi *NaharPariyojana* during 1981-1991

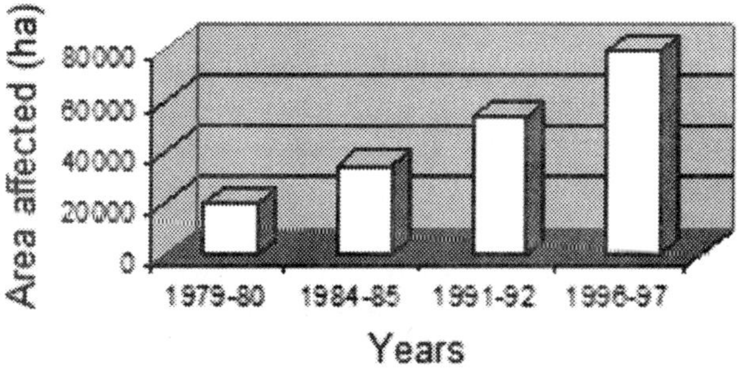

Fig. 1.3 Area affected by water logging and soil salinity in Upper Krishna Project, Karnataka

Water Logging: Definition and Criteria

In the simplest form water logging connotes restricted drainage, whether surface or subsurface. `Clayton' has defined a land to be water-logged when the water table is within ± 45.75 cm of the ground surface. The definition was first adopted in India, by the Drainage Board, in 1917. A modification to this definition was made in 1925 in the state of Punjab, when a land was designated as waterlogged, if it is thrown permanently out of cultivation due to rise in sub soil water and is recorded as *banjar* in the revenue records. The water table in the month of June should be within 1.25 m of the surface. A further modification in the definition was made to account for the yield of crops. It was decided to term a land to be waterlogged if the yield is reduced to 25% of the optimum value, due to rise in sub-soil water. National Commission on Agriculture (1976) accepted the definition given by a committee of Central Water and Power Commission. According to this definition `a soil is waterlogged if the water table rises to an extent that root zone becomes saturated, diffusion of air is curtailed and amount of oxygen is reduced with increase in carbon dioxide'. The criterion given by Clayton is commonly used in India and abroad due to its simplicity. However, present day drainage definition apart from extraction of excess water includes improvement of the water, air and temperature conditions in soils. It is now realized that differentiation be made in water logging due to water stagnation (land submergence) and water logging due to rise in the water table because the strategies for the reclamation and management of the two groups of soil would differ. Different states follow different norms for defining water logging (Table 1.5). Therefore, there may be some discrepancies in the estimates of the areas made by different agencies.

Working Group (1991) set-up by the Ministry of Water Resources adopted the definition given by Central Water and Power Commission. It is mentioned that the water table which is considered harmful would depend upon the type of crop, type of the soil and the quality of water. The actual depth to water table, when it adversely affects the crops yield, may vary over a wide range from zero for rice to about 1.5 m for other crops. Following norms are suggested.

Depth to water table (m)	*Nomenclature suggested*
< 2	Waterlogged
2-3	Potentially waterlogged
> 3	Safe

Table 1.5 Norms for categorization of waterlogged areas in states

State	Water table depth below the ground level (m)
Uttar Pradesh (Sharda Sahayak Project)	
Worst zone	Less than 1
Bad zone	1- 2
Alarming zone	2-3
Safe zone	More than 3
Punjab	
Very critical	0-1.5
Critical	0 -2
Haryana	
Critical waterlogged	0-1.5
Critical	1.5-3
Karnataka	
Tungbhadra command	0-2
Himachal Pradesh	0-2
Maharashtra	
Fully waterlogged	Water at surface
Waterlogged	0-1.2

Causes of Water Logging

Water logging is caused both by natural and manmade factors. Some of the factors responsible for water logging are as follows:

- Poor natural drainage as a consequence of unfavorable topography or unfavorable sub-soil geology like existence of hardpan at shallow depths
- Heavy storm rainfall coupled with poor natural drainage
- Spilling of rivers resulting in submergence of agricultural lands
- Introduction of irrigation without taking into account characteristics of soils and sub-soils for their irrigability
- Heavy losses of water due to seepage from canals, distributaries and watercourses

- Poor on-farm water management resulting in poor application efficiencies
- Development activities such as construction of roads, bridges, railway lines and buildings resulting in choking of natural drainage
- Poor maintenance of existing drainage system and outlets
- Deforestation and poor upkeep of watersheds
- Hydraulic pressure of water from upper irrigated areas resulting in seepage outcrop in low lying areas.

Irrigation Induced Water Logging

The water logging in irrigation commands could appear in the form of surface stagnation or shallow water table.

Surface stagnation of water

Surface stagnation of water is a very common problem in the monsoon climatic conditions and is experienced in almost all the irrigation commands. Analysis of rainfall data for various stations in the country revealed that such a possibility cannot be ruled out even in semiarid or arid climatic conditions. With introduction of irrigation, degree of water stagnation and the periods for which water stagnates on the land surface goes a sea change. It is because capacity of the soil to absorb water reduces due to rise in the water table. The developmental activities like the construction of canal net work, roads and railway lines in the irrigation project also obstruct the natural drainage to increase degree and duration of water stagnation and recharge to the ground water. Thus, it is necessary to plan for surface drainage improvements whenever irrigation projects are implemented.

Shallow water table

India has a long history of environmental problems arising from the introduction of irrigation. The first recorded incidence of water logging / salinity is in 1855 in the Western Yamuna Canal. Soil and water samples were taken to London for analysis in the Royal School of Mines. The analysis suggested a practical remedy that drainage should accompany irrigation. Incidences of alkali development were also reported from areas around Aligarh in the command of Ganga canal. Reports of salinity problems also came from the western state of Maharashtra where a reclamation program was undertaken by Inglish and Gokhale (1927). In the state of Punjab soil surveys, which were locally termed as *Thurgirdawari* were initiated. These surveys evaluated the extent of damage caused by salinity (*Thur*) or high moisture content (*Sem*) in the root zone and were based on visual observations. Even in this century, so spectacular has been the rise of water table in some of the irrigated lands that used to be under dry farming only a few years back, have become unfit for cultivation of any upland crop during the rainy season. For example, parts of Rohtak district in Haryana where water table was more than 20 m deep experienced a rise of water table at the rate of 0.6 m/annum reaching an average depth of 3.0 m in 1982. At Haryana Agricultural University farm, where the water table used to be at 15.6 m in 1967 at the time of introduction of irrigation through Bhakra canal system rose to 2.0 m in 1984 (Kumar and Singh, 1994). In the Ukai Kakrapar project where first stage was completed in 1957-58 and second stage in 1972, the area where water table was within 1.5 m of the soil surface during the post monsoon period went up to 29,971 ha in 1990 (Table 1.6). The area having water table within 3 m was 132,276 ha.

Table 1.6 Area under various hydro isobaths (water table below soil surface) during post monsoon season in Kakrapar project, Gujarat (ha)

Period	Depth of water table (m)				
	0-1.5	1.5-3.0	3.0-6.0	6.0-9.0	>9.0
1957-58	0.0	1036	49663	195148	160152
1989-90	29971.0	102305	147997	55425	*

*Area included in 6.0-9.0 m range

Note: The area during 1989-90 does not include *Khar lands* and non- project command (NPC) which were included in pre- irrigation survey.

Even in the driest part of the country where irrigation was introduced in 1961 through Indira Gandhi *Nahar Pariyojana* (IGNP) in Rajasthan, a steep rise in water table was reported varying from 0.6 m/annum in Gang canal to 1.0 m/annum in IGNP (Hooja *et al.*, 1994). The area irrigated through lift irrigation schemes was also affected with this malady although only limited quantities of water were introduced through these schemes. In the command area of Jui lift irrigation canal in Haryana, there has been continuous rise of water table ranging from 0.3 to 0.8 m per annum (Sharma and Bhargava, 1982). The compilation of data by Singh (1993) confirms that once canal irrigation is introduced, the water table rise is inevitable (Table 1.7).

Table 1.7 Rate of rise in water table in different irrigation commands

Irrigation command	Rate of rise in water table (m/annum)
Mahi Right Bank Canal Command (MRBC), Gujarat	0.28
Rajasthan Canal Command (now IGNP), Rajasthan	0.29-0.88
Western Yamuna & Bhakra Canal Commands, Haryana	0.30-1.00
Sirhind Canal Command, Punjab	0.10-1.00
Sharda Sahayak Canal Command, Uttar Pradesh	0.68
Malprabha Canal Command, Karnataka	0.60-1.2
Nagarujana Sagar Irrigation Project, Andhra Pradesh	0.32
Sriram Sagar Irrigation Project, Andhra Pradesh	0.26

Measurement of Water Table

To assess the degree of water logging and/or to classify waterlogged lands, one of the essential pre-requisites is to measure depth of water table. Commonly adopted methodologies for this purpose are given in the following paragraphs.

Water table in open wells

Open well is a very common feature of Indian rural landscape. These wells have been an important source of water to meet the domestic and agricultural water requirement. The water table in these wells at a particular time of the year and in different seasons will reveal the ground water table behavior in the region. Central Ground Water Board/ State Directorate of Ground Water have fully utilized the potential of these wells in monitoring ground water table behavior. Usually pre monsoon and post monsoon ground water table is measured by these departments. Recently, few ideally located wells have also been earmarked to undertake monthly reading and to assess water quality. These data provide wealth of information to researchers and field functionaries. Sometimes, existing wells are concentrated at few points while in others the density of wells is quite sparse. It poses a major limitation in the use of these wells to arrive at a realistic field situation. Thus, there is a need to install more observation wells/piezometers to minimize gaps in information and to have a better insight into the regional problem. A rough guide to decide the number of wells/observation wells is given in Table 1.8.

Table 1.8 Wells/observation wells required in relation to the size of the drainage area

Area (ha)	No. of wells/observation wells	No. per 100 ha
100	20	20
1000	40	4
10000	100	4
100000	300	0.3

Installation of observation wells/piezometers

The simplest form of an observation well (though of temporary nature) is an auger hole left uncased in the field which may be employed to measure water table depth during an exploratory survey. A permanent type observation well consists of a pipe that is perforated from bottom to about 50 cm below the ground surface (Fig. 1.4). The perforations in an observation well are drilled in five to six rows and are covered with jute/coir rope or muslin cloth to prevent fine particles from entering the pipe or clogging the tiny holes. The lower open end of the pipe is sealed with a plug. The pipe is placed in the centre of a bore hole made with a hand or power auger and the annular space between the pipe and wall of the hole over the entire length of the perforated portion is filled with gravel pack or coarse sand. The gravel pack facilitates the flow of water from the soil into the pipe and vice versa. The rest of the annular space is filled with fine materials (clay, clayey fine sand) whose permeability is low enough to prevent leakage along the pipe. To keep the pipe in position and to prevent any direct contact between surface water and the backfill material, a concrete seal is provided around the pipe, extending from the ground surface to a depth of about 15 cm. The top of the pipe is fitted with a cap to prevent entry of foreign materials into the pipe. The water level in an observation well represents the average hydraulic head over the length of the perforated portion of the pipe. At shallow depths, this head corresponds to the ground water table.

A piezometer is an open ended non-perforated galvanized iron (or sometimes plastic) pipe of 3 to 5 cm diameter, driven or jetted into the ground to measure the hydraulic head. To ensure that the piezometer functions properly, the lower 0.5 m of the pipe is perforated (Fig. 1.5). The water level in the pipe corresponds with the hydraulic head at the lower end where the pipe terminates. Although both observation wells and piezometers are required in the initial stages, piezometers are installed selectively. Piezometers are used to measure the hydrostatic pressure at different depths in the sub soil layers. As such, a battery of piezometers is installed at one location with the lower end of each piezometer terminating at different depths. The number is increased if the problem of artisan pressure is indicated.

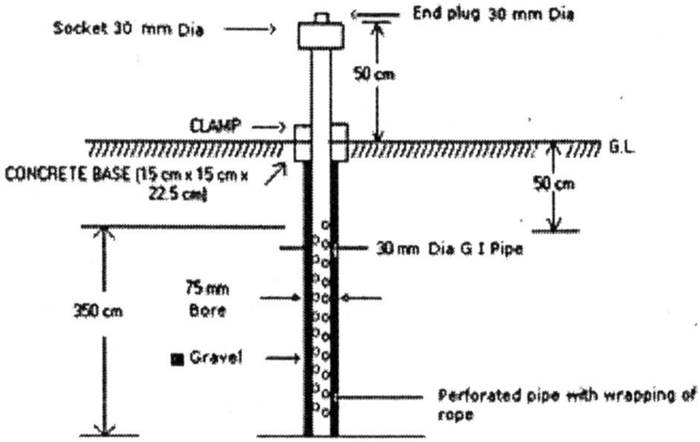

Fig. 1.4 A cross-sectional view of a cased observation well

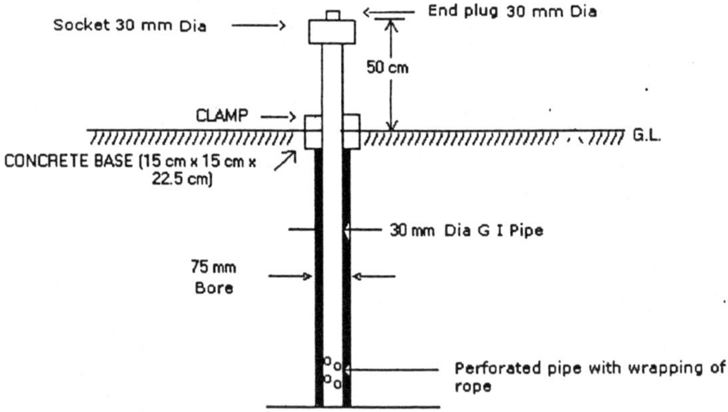

Fig. 1.5 Schematic cross-section through a piezometer

Measurement equipment

The most commonly time tested method of chain and bell sounding device fitted at the bottom end of a thin steel or metallic tape would be a perfect arrangement for measuring depth to water table as well as drawing the water sample. The device is a 1-1.5 cm galvanized iron pipe of about 10-20 cm length. The bottom end is sealed with a rubber cork leaving about a 1 mm gap (Fig. 1.6). Arrangement for attaching tape is made at the upper end. When lowered into the well, it makes a clearly audible "plop" sound as it contacts the water surface. Once the water table measurement is over, the device is lowered, given few up and down jerks to mix the water and sample is drawn. It is advisable to throw the first sample as it may not be representative of the ground water. The equipment can be prepared locally and being inexpensive, a number of them could be prepared and kept handy for use.

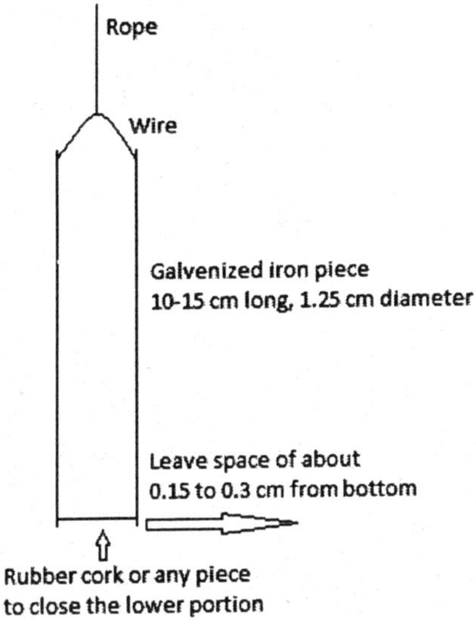

Fig. 1.6 Chain and bell sounding device to measure depth to water table

Extent of Water Logging

The precise estimate of the extent of water logging is not available. There are no universally accepted criteria to characterize a land as waterlogged. Attempts have been made to estimate the waterlogged area by different agencies at different points of time. About 40 M ha is estimated to be flood prone. On an average basis, about 8 M ha agricultural land is affected by floods annually. The National Commission on Agriculture (1976) estimated that waterlogged areas cover about 6 M ha. The data includes water logging both due to surface stagnation of water and rise in water table in irrigation commands and the humid regions of the country. This estimate appears to be most appropriate. Out of 6 M ha area affected by water logging, 3.4 M ha is due to surface flooding and 2.6 M ha due to rise in water table. The last figure closely matches with the recent estimate of the Working Group constituted to identify waterlogged areas due to rise in the water table in different irrigation commands (Table 1.9).

Table 1.9 Extent of waterlogged area as estimated by various agencies (in lakh ha)

State	As per National Commission on Agriculture (1976)	As per reconciled estimate of the Working Group (1991)
Andhra Pradesh	3.39	2.66
Assam	N.R	N.R
Bihar	1.17	6.20
Gujarat	4.84	1.72
Haryana	6.20	2.49
Jammu & Kashmir	0.10	0.01
Karnataka	0.10	0.24
Kerala	0.61	0.12
Madhya Pradesh	0.57	0.73
Maharashtra	1.11	0.15
Odisha	0.60	1.96
Punjab	10.90	2.00
Rajasthan	3.48	1.80
Tamil Nadu	0.18	0.16
Uttar Pradesh	8.10	4.30
West Bengal	18.50	N.R
Delhi	0.01	N.R
Total (lakh ha)	**59.86**	**24.56**
Total (M ha)	**6.00**	**2.46**

Source: Working Group (1991), N.R= Not reported

IRRIGATION INDUCED SOIL SALINITY

Soil salinity problems in irrigation commands develop whenever soil and hydrological conditions favor the accumulation of soluble salts in the root zone. The rise in water table in semiarid and arid areas mobilizes the salts present in the soil profile and ground water. Once the ground water table rises between 2 and 3 m of the soil surface, it contributes substantially to evaporation from the soil surface and water uptake by plants. The upward flux of water due to evaporation and water uptake by plants results in gradual concentration of salts in the root zone (Fig. 1.7 left and right). In the initial years, the crop yield may be reduced due to salinity, but as the severity of the problem increases, the lands may have to be abandoned since the excess salts render the soils completely unproductive.

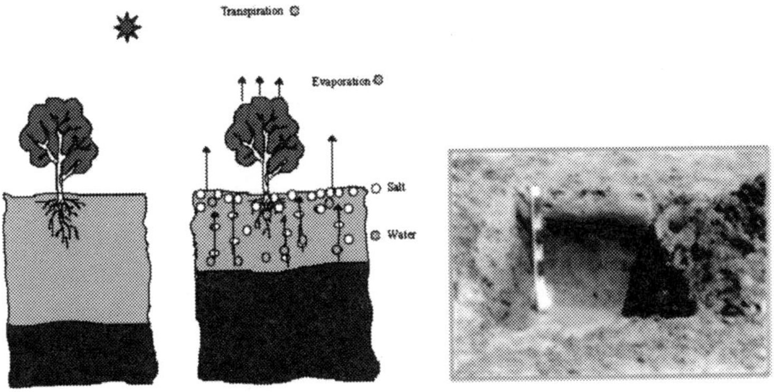

Fig. 1.7 Capillary rise and soil salinization in areas having shallow saline water table (left) schematic and (right) photo from Ukai-Karapar project, Gujarat

Causes of Soil Salinization

Salt affected soils may develop from normal soils through the accumulation of salts in one of the following ways. A combination of these processes could be operative at several places.

- Accumulation of salts applied with irrigation water
- Movement of salts from lower layers
- Upward movement of ground water through capillary rise
- Sea water inundation in coastal regions
- Deposition of windblown salts.

Extent of Salt Affected Soils

The first estimate of the Central Soil Salinity Research Institute, Karnal made in 1970 placed the extent of salt assessed soils at 7.0 M ha. It was revised to 8.5 M ha (Singh, 1992). Current estimate is 6.73 M ha (Table 1.10). It is to be noted that the area reported in this table is for soils that have already gone out of cultivation. There could be large areas where salinity problem has tended to affect crop yields. As per the future projections made on an all India basis, an area of about 11 M ha is likely to be affected by the problems of water logging and soil salinity in the irrigation commands of India by 2025 (CSSRI, 1997).

Table 1.10 Extent of salt affected soils in India (ha)

State	Saline	Sodic	Total
Andhra Pradesh	77598	196609	274207
Andaman & Nicobar Island	77000	0	77000
Bihar	47301	105852	153153
Gujarat	1680570	541430	2222000
Haryana	49157	183399	232556
Karnataka	1893	148136	150029
Kerala	20000	0	20000
Madhya Pradesh	0	139720	139720
Maharashtra	184089	422670	606759
Odisha	147138	0	147138
Punjab	0	151717	151717
Rajasthan	195571	179371	374942
Tamil Nadu	13231	354784	368015
Uttar Pradesh	21989	1346971	1368960
West Bengal	441272	0	441272
Total (ha)	**2956809**	**3770659**	**6727468**
Total (M ha)	**2.96**	**3.77**	**6.73**

Source: Mondal *et al.* (2011)

Types of Salinity Problem

Conventionally, salt affected soils have been grouped as alkali, saline and saline-alkali soils (Table 1.11). Considering specific management requirements, these soils have been categorized in two main classes namely (i) alkali, and (ii) saline soils (Fig. 1.8 left and Fig. 1.8 right).

Alkali soils

These soils contain excess of salts like bicarbonates, carbonates and silicate of sodium, capable of alkaline hydrolysis or have sufficient exchangeable sodium to interfere with growth of most crops. They have pHs greater than 8.5, electrolyte conductivity of saturation extract less than 4 dS/m and exchangeable sodium percentage (ESP) of 15 or more. It is estimated that alkali (sodic) soils cover an area of 3.77 M ha in alluvial and black soils regions mainly in Punjab, Haryana, Uttar Pradesh and Madhya Pradesh (Table 1.10).

Table 1.11 Classification of salt affected soils on the basis of chemical characteristics

Class	EC_e(dS/m)	ESP[§]	pH_2[$]	Indian Local Names
Saline Soils	> 4*	< 15	< 8.5	*Thur, Uippu, Lona, Shora, Soula, Pokhali, Khar and Kari*
Saline-alkali soils	> 4	>15	Variable	*Usar, Kallar, Karl, Chopan, Bari, Reh, Choudu and Kshar*
Alkali	< 4	>15	8.5-10.0	*Usar, Rakkar, Bara, Chopan, Kari*

*Present literature sometimes gives this value as 2.0 dS/m. [§] Sometimes SAR is used instead of ESP. For Indian conditions, ESP is preferable. If pH_s is measured, the value is taken as 8.2.

Saline soils

These soils contain neutral salts (chlorides and sulphates of sodium, calcium and magnesium) in quantities enough to interfere with growth of most crop plants. The pH of saturated soil paste is less than 8.5, Soil EC_e is more than 4 dS/m and ESP less than 15. Currently, the criterion has been revised and a soil having EC_e more than 2 dS/m is characterized as saline soil, mainly because many sensitive crops such as gram and vegetables start suffering at this salinity. Problem of soil salinity is widespread in the arid and semiarid regions (irrigated lands) and sub-humid to humid regions (coastal salinity). A total of 2.96 M ha area is affected by soil salinity which is quite widespread in alluvial, black, red and grey soils (Table 1.10). Following three distinct groups of saline soils could be identified.

- Inland saline soils having shallow water table, water quality being saline in nature. These soils suffer from water logging and salinity.
- Normal soils being irrigated over the years with saline and/or alkali ground water. These soils normally have deep water tablc. However, in certain situations the water table could be shallow. Theses soils generally occur in arid and semiarid regions.
- Coastal saline soils where the problem is due to ingress of sea water resulting in high salinity and temporary flooding. The climate is generally humid to semiarid/arid.

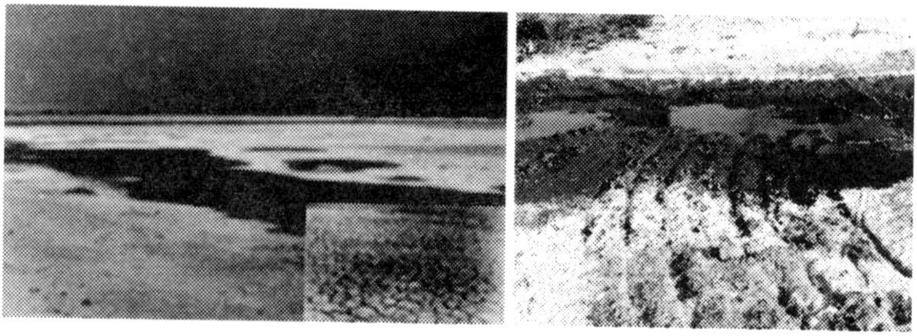

Fig. 1.8 A view of alkali land (left) and a saline land (right) in the Indo-Gangetic plain

Saline alkali soils

Soils having both high EC_e and ESP have been categorized as saline-alkali soils. However, from reclamation/management point of view, these soils would either be reclaimed as saline or alkali lands. As such, Central Soil Salinity Research Institute has categorized the soils either as saline or alkali. The following guidelines have been given to categorize saline-alkali soils as saline or alkali.

Soils category	Guidelines
Saline	In the saturated paste extract, the ratio of soluble ions should be $(CO_3 + HCO_3) / (Cl + SO_4) < 1$, $Na/ (Cl + SO_4) < 1$
Alkali	In the saturated paste extract, the ratio of soluble ions should be $(CO_3 + HCO_3) / (Cl + SO_4) > 1$, $Na/ (Cl + SO_4) > 1$

LABORATORY INVESTIGATIONS FOR DIAGNOSIS

Electrical Conductivity (EC)

The total salt content of the soil solution or of the irrigation water is represented in 3 ways i.e. milligrams per liter (mg/L), millimole per liter (mmol/L) or electrolyte conductivity expressed in dS/m. Although EC has the units of Siemens/m, it is commonly reported in deci Siemens/m (dS/m) which is equivalent to the earlier unit expressed in millimhos/cm (mmhos/cm) (1 dS/m = 1 mmhos/cm). Milliequivalent per liter is the milliequivalent weight of the anion or compound (1/1000 part of the equivalent weight) in a liter of solution. Electrolyte conductivity is by far the simplest representation and is a measure of the ease with which current will flow through the soil water or the irrigation water. All the 3 units are related with each other. The relation between mmol/L and EC is

$$\text{mmol/L} = 10 \times \text{EC} \qquad 0.1<\text{EC}<5.0 \tag{1.1}$$

Relation between mg/L and EC is

$$\text{mg/L} = 0.64 \times \text{EC} \qquad 0.1<\text{EC}<5.0 \tag{1.2}$$

The value of the factor in eq. (1.2) increases with increasing EC (greater than 5 dS/m), and tends to be near unity at very high salinity levels. These relations are only approximate and have been obtained by plotting the values for several salts commonly occurring in the soils. Ionic composition in mmol/L is converted to mg/L by multiplying with its equivalent weight.

Electrolyte conductivity has the advantage of simplicity in its measurement as it can be readily measured in the field or laboratory by using a conductivity meter. It is closely related to the sum of cations (or anions) as determined chemically and usually correlates closely with the total dissolved solids. As the electrolyte conductivity would depend upon the moisture content and temperature; it is common to express the conductivity at common moisture and temperature. The total soluble salt concentration is estimated by measuring the EC of the saturation extract at 25°C. Most conductivity meters are provided with temperature correction. If such correction is not provided, electrolyte conductivity so determined is corrected for temperature.

Conductivity meters are usually provided with an immersion type conductivity cell with platinum electrodes. Following stepwise procedure is used to measure EC.

- Prepare a 0.01 N potassium chloride solution by dissolving 0.7456 g of potassium chloride in distilled water and make it to 1 L at 25°C
- To calibrate the electrode, measure the EC of the standard solution at 25°C. This solution at 25°C has an electrolyte conductivity of 1.412 dS/m
- Determine the cell constant by the following equation

$$\text{Cell constant} = \frac{\text{EC of the standard solution at } 25^\circ\text{C } (1.412 \text{ dS/m})}{\text{Measured EC of the standard solution at } 25^\circ\text{C}} \tag{1.3}$$

- Determine the EC of the soil saturation extract or water
- Multiply it with cell constant determined using eq. (1.3)
- Measure the temperature of the sample, if temperature correction is not provided in the EC meter. Correct the EC

value to 25°C by multiplying it with temperature correction factor (Table 1.12). As a thumb rule, the electrical conductivity increases approximately by 2% per degree C rise in temperature. For example to convert values measured at 20°C to the values at 25°C, multiply by 1.112. However, ft values from the table could be used as follows:

$$\text{EC at } 25^{\circ}\text{C} = \text{EC at } T^{\circ}\text{C} \times f_t$$

Here, f_t = correction factor at temperature t from Table 1.12.

In many soil testing laboratories for quick appraisal of soils, EC is usually determined in 1:2 soil water extract (EC_2) which is prepared by dissolving 10 g of dry soil with 20 g of water. For heavy textured soil even 1:5 (EC_5) soil water extracts have been used. EC_2 for medium alluvial soils is highly and significantly correlated with EC_e (Gupta *et al.*, 1965; Singhla *et al.*, 1974). A calibration curve relating EC_2 or EC_5 and EC_e is used to estimate EC_e. A typical curve used for this purpose for a medium textured soil is given in Fig. 1.9. It shows that for the soil used in calibration, EC_e is nearly 4 times the EC_2. Following relation could also be used to calculate the EC_e.

$$EC_e = \frac{\theta_2}{\theta_{sp}} * EC_2 \tag{1.4}$$

Here θ_2 and θ_{sp} are the soil moisture contents in soil-water paste and at saturation for a given soil. EC of the soil solution at field capacity (EC_{fc}) is usually taken as twice that of EC_e.

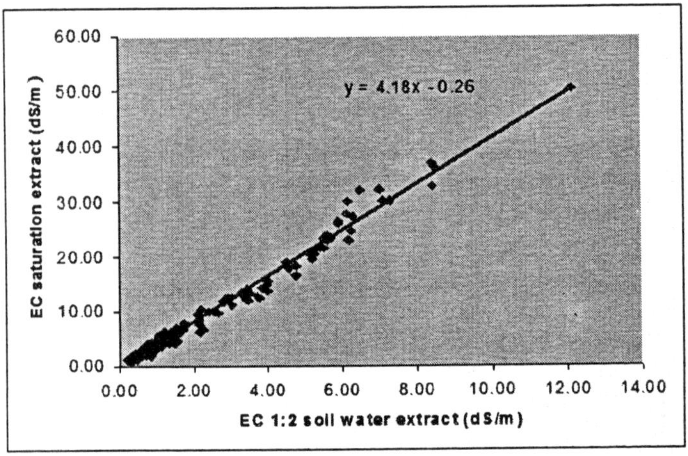

Fig. 1.9 Relationship between EC_e and EC_2 for a medium textured soil

Table 1.12 Correction factors for conductivity data at standard temperature of 25°C

°C	f_t	°C	f_t	°C	f_t
08.0	1.488	22.6	1.051	29.2	0.921
09.0	1.448	22.8	1.047	29.4	0.918
10.0	1.411	23.0	1.043	29.6	0.914
11.0	1.375	23.2	1.038	29.8	0.911
12.0	1.341	23.4	1.034	30.0	0.907
13.0	1.309	23.6	1.029	30.2	0.904
14.0	1.277	23.8	1.025	30.4	0.901
15.0	1.247	24.0	1.020	30.6	0.897
16.0	1.218	24.2	1.016	30.8	0.894
17.0	1.189	24.4	1.012	31.0	0.890
18.0	1.163	24.6	1.008	31.2	0.887
18.2	1.157	24.8	1.004	31.4	0.884
18.4	1.152	25.0	1.000	31.6	0.880
18.6	1.147	25.2	0.996	31.8	0.877
18.8	1.142	25.4	0.992	32.0	0.873
19.0	1.136	25.6	0.988	32.2	0.870
19.2	1.131	25.8	0.983	32.4	0.867
19.4	1.127	26.0	0.979	32.6	0.864
19.6	1.122	26.2	0.975	32.8	0.861
19.8	1.117	26.4	0.971	33.0	0.858
20.0	1.112	26.6	0.967	34.0	0.843
20.2	1.107	26.8	0.964	35.0	0.829
20.4	1.102	27.0	0.960	36.0	0.815
20.6	1.097	27.2	0.956	37.0	0.801
20.8	1.092	27.4	0.953	380	0.788
21.0	1.087	27.6	0.950	39.0	0.755
21.2	1.082	27.8	0.947	40.0	0.763
21.4	1.078	28.0	0.943	41.0	0.750
21.6	1.073	28.2	0.940	42.0	0.739
21.8	1.068	28.4	0.936	43.0	0.727
22.0	1.064	28.6	0.932	44.0	0.716
22.2	1.060	28.8	0.929	45.0	0.705
22.4	1.055	29.0	0.925	46.0	0.694

Source: USSLS (1954)

Soil Reaction, pH

The pH scale is a measure of the effective concentration of the hydrogen ion. pH is the log of the inverse of the hydrogen ion concentration in the saturated paste of the soil (pH_s). It is used as an index of the acidity or the alkalinity of the soil. A pH of 7.0 is neutral. Soils are acidic below pH 7.0 and alkaline above pH 7.0. A pH_s of 8.2 or above characterizes soil as alkali. When rapid determinations are needed, soil water suspensions (1:2) are used to determine pH.

A pH meter with glass electrode is used to measure the pH of the saturated soil paste (pH_s) or pH of the 1:2 soil water extract (pH_2). Soil pH_2 and pH_s are significantly correlated (Gupta, 1972). It is desirable to use the same procedure in all determinations so that the values are consistent and have maximum diagnostic value. The pH besides being a diagnostic tool, has been used to evaluate gypsum requirement since pH of the alkali soils of the Indo-Gangetic alluvial plains is correlated well with the ESP. Following step wise procedure is used to measure pH.

- Calibrate the pH meter using a standard pH solution by dissolving standard pH buffer tablets of pH 7.0 and 9.2
- Use the same soil saturation extract/water sample as used in EC determination. Soil water extract or even paste in 1:2 or 1:5 could be used
- Immerse the electrodes of the pH meter in the sample
- Measure the pH of the sample

It is always useful to test the calibration of the pH meter in between if large numbers of samples are being analyzed.

Table 1.13 Relationship between pH_s, pH_2 and ESP of the alluvial soils

pH_s	pH_2	ESP
8.0-8.2	8.5-9.2	< 20
8.2-8.4	9.2-9.4	20-35
8.4-8.6	9.4-9.7	35-50
8.6-8.8	9.7-9.8	50-65
8.8-9.0	9.8-9.9	65-75
9.0-9.2	9.9-10.6	85-100

Like EC, pH_2 and pH_s are also related, pH_2 being higher than pH_s. The difference increases with increase in pH_s (Table 1.13, Gupta, 1972).

It may be noted that in alluvial soils of the Indo-Gangetic plains, there is a close relationship between the pH_s, pH_2 and ESP. Unfortunately, this relation does not hold true in many soils and as such, ESP needs to determined to calculate gypsum requirement of such soils.

Exchangeable sodium percentage (ESP)

Each soil has certain capacity to adsorb cations (the positively charged constituents of the dissolved salts) such as calcium (Ca^{++}), magnesium (Mg^{++}), sodium (Na^{+}) and potassium (K^{+}). It is known as cation exchange capacity (CEC) and is expressed as c mol (p^{+})/kg of soil. Depending upon the concentration of these ions in the soil solution, they get adsorbed. Once the relative concentration of these ions changes in the soil solution, the cations could be replaced by each other although the capacity of the soil remains the same. Exchangeable sodium percentage is the amount of adsorbed sodium, expressed in percent of the CEC of the soil.

$$ESP = \frac{\text{Sodium on the exchange complex (c mol (p}^{+}\text{) / kg soil)}}{\text{CEC (c mol (p}^{+}\text{) / kg soil)}} \times 100 \qquad (1.5)$$

Exchangeable Cations and Cation Exchange Capacity

Reagents

- **Alcoholic ammonium chloride, 1M:** Dissolve 53.5 g NH_4Cl in about 300 ml of water and add 600 ml of ethanol (95% v/v). Adjust the pH to 8.5-8.6 with concentrated ammonia solution and make up to 1 L. Store in a well stoppard bottle to prevent access of CO_2.
- **Ammonium ethanoate solution, 1M:** Dilute approximately 230 ml of glacial ethanoic (acetic) acid to 1 L. Dilute approximately 220 ml of ammonium solution (approximately 35% m/m NH_3) to 1 L. Mix the solutions together and adjust to pH 7 with ethanoic acid or ammonia. Dilute to 4 L.
- **Ethanol, 95 per cent**
- **Potassium chloride solution:** Dissolve approximately 100 g KCl in water and make to 1 L

Procedure

1. Weigh 5 g (± 0.01 g) of < 2 mm air dry soil into a 100 ml beaker. Add 20 ml of ammonium ethanoate solution. Stir and allow to stand overnight. Transfer the suspension to a filter funnel fitted with a Whatman no. 44 filter paper standing in a 250 ml volumetric flask. Thoroughly wash the beaker with ammonium ethanoate solution into the funnel. Leach the soil with successive 25 ml volume of ammonium ethanoate solution. Continue until 250 ml of filtrate is collected. Make up to the mark with ammonium ethanoate solution.
2. Use the extract so collected to determine potassium, calcium and magnesium.
3. Follow the same procedure with alcoholic ammonium chloride and use the leachate to determine sodium.
4. Sodium so determined minus the soluble sodium present in the soil saturation extract would give the exchangeable sodium.

Since, the exchange capacity of the saline and sodic soils is dominated by Na^+, Ca^{2+} and Mg^{2+} with small amounts of K^+, once these are known, the CEC could be determined by adding the exchangeable cations and ESP using eq. (1.5). More accurate procedures are available in standard books but for practical land reclamation, such approximate analysis would not pose any difficulties (Gupta *et al.*, 2012).

Sodium Adsorption Ratio, SAR

Ratio of the monovalent and the divalent cations is used as an index of the exchangeable sodium or ESP. As the potassium on the exchangeable complex is comparatively less than the sodium, it is neglected in the calculations. On this basis, an index sodium adsorption ratio is given as

$$SAR = \frac{Na}{\sqrt{(Ca + Mg/2)}} \tag{1.6}$$

Here concentrations of ions in the soil solution are in meq/L. If concentrations are in mmol/L, then SAR is given as

$$SAR = \frac{Na}{\sqrt{(Ca + Mg)}} \quad (mmol/L)^{1/2} \tag{1.7}$$

Due to different behavior of Mg than Ca, it has been suggested to calculate SAR as $\frac{Na}{\sqrt{Ca}}$ (Gupta, 1984; Keren, 2000)

Residual Sodium Carbonate (RSC)

Eaton (1950) pointed out that if irrigation water has an appreciable carbonate and bicarbonate content, calcium and magnesium carbonate will precipitate when the concentration of the soil solution is increased through evapotranspiration. On the assumption that their precipitation was complete, he proposed the concept of residual sodium carbonate (RSC) for evaluating the quality of bicarbonate containing water and defined it in meq/L as:

$$RSC = (CO_3^{2-} + HCO_3^{1-}) - (Ca^{2+} + Mg^{2+}) \quad (1.8)$$

US Salinity Laboratory Staff (1954) recommended that both RSC and SAR be used to evaluate the sodicity hazard. Subsequent investigators showed that general reliability of the RSC concept for evaluating the sodium hazard of bicarbonate waters is questionable, mainly because of the assumption of quantitative precipitation (Babcock *et al.*, 1973). Nonetheless, it continues to be a good indicator of the alkali hazard of a particular water as the SAR of the water would tend to be much more than anticipated due to concentration of the soil solution alone.

DIAGNOSIS OF SALINE/ALKALI SOILS

Some visual observations during soil surveys could reveal the nature of the problem. The visual characters that serve as a diagnostic criterion in identifying the two groups of soil are reported in Table 1.14. Exact analysis should however, be based on the chemical characteristics as revealed by the sample analysis. The differentiating parameters have already been described in Table 1.11.

Based on the ESP and EC_e of the soils, the hazard on crop productivity has been graded from slight or negligible to extremely high or severe (Table 1.15 and Table 1.16).

Salt Affected Soil Series: An Overview

Soils in India have been categorized in 64 benchmark soils out of which 9 have been found to be salt affected. To illustrate general characteristics, a soil series affected by alkali salts is described in Table 1.17. The characteristics of a saline soil profile are given in Table 1.18. It is apparent that alkali soils have high pH and high ESP. The pH is more than 9 in all the layers and ESP exceeds 85 at least in the top layers of the soil profile. Although, the electrical conductivity of the alkali soil is also high at least in the top 64 cm, a chemical amendment is required to reclaim this land (Table 1.17).

Table 1.14 The visual observations to distinguish an alkali and a saline soil

Observation	Alkali soil	Saline soil
Rain water stagnation	Stagnation for longer period	Relatively lesser period
Stagnated water type	Dark coloured, muddy, soapy	Clear
Surface salt crust	Dark/brown/ash coloured clay crust, black spots could be seen	White salt efflorescence
Ground water	Usually good, occasionally with high RSC	Saline, sometimes with high SAR
Presence of nodules	$CaCO_3$ nodules are present	Gypsum may be present at some depth
Natural vegetation (grasses)	*Sporobolus marginatus, Desmostachya bipinata, Suaeda maritma, Diplachne fusca, Dichanthium annulatum, Cynodon dactylon, Chloris barbata, Brachiaria mutica, Kochia indica Panicum antidotale*	*Cressacretica, Cyperus rotundus, Chloris pallida, Sporobolus pallidus, Haloxylonsali cornicum, Acluropus lagopoides, Zygophyllum simplex, Dichanthium annulatum, Sueda furuticosa, Butea monosperma*

Table 1.15 ESP and degree of alkali hazard

Approximate ESP	Alkali hazard
Up to 15	None to slight
15 – 30	Slight to moderate
30 – 50	Moderate to high
50 – 70	High to very high
> 70	Extremely high

Table 1.16 EC_e and degree of salinity hazard

Soil salinity class	EC_e (dS/m)	Effect on crop plants
Non-saline	0 - 2	Salinity effect negligible
Slightly saline	2 – 4	Yields of sensitive crops restricted
Moderately saline	4 – 8	Yields of many crops restricted
Strongly saline	8 – 16	Tolerant crops give satisfactory yields
Very strongly saline	> 16	Only a few very tolerant crops yield satisfactorily

Table 1.17 Physico-chemical characteristics of an alkali soil from Etah (U.P.) representing Indo-Gangetic plain

Depth (cm)	EC_e (dS/m)	pH_s	Ionic composition (meq/L)						
			Ca+Mg	Na	K	CO_3	HCO_3	Cl	SO_4
0-16	13.84	10.2	2.2	210.0	0.2	151.0	28.0	14.0	21.0
16-27	12.50	10.1	2.5	190.0	0.2	129.0	20.0	14.0	13.0
27-64	4.60	9.7	1.3	45.0	0.1	32.0	10.0	11.0	3.0
64-81	1.92	9.8	2.5	19.0	0.1	4.0	13.0	5.0	1.0
81-108	1.88	9.4	2.0	16.0	0.1	4.0	9.0	6.0	1.0
108-127	2.35	9.1	1.0	21.0	0.1	7.0	10.0	6.0	1.0
127-148	1.40	8.6	2.7	13.0	0.1	2.0	6.0	6.0	1.0

Depth (cm)	Sand (%)	Silt (%)	Clay (%)	Textural Class* (%)	$CaCO_3$ (%)	OM** (%)	CEC cmol (p^+)/kg	Soil ESP
0-16	62.0	24.0	14.0	sl	1.9	0.7	6.7	88
16-27	57.0	25.5	17.5	sl	4.0	0.6	7.3	87
27-64	45.0	22.0	33.0	scl	3.8	0.6	15.9	67
64-81	38.0	35.0	27.0	l	2.7	0.6	13.9	78
81-108	34.0	34.0	32.0	cl	12.0	0.5	12.9	24
108-127	39.5	33.0	27.5	c	23.8	0.5	12.0	31
127-148	65.0	18.0	17.0	sl	25.0	0.4	8.0	22

* sl sandy loam, scl sandy clay loam, l loam, cl clay loam, c clay,
** OM organic matter, CEC cation exchange capacity

NEED TO RECLAIM SALT AFFECTED SOILS

Land and water resources in agriculture are rapidly shrinking because of keen competition between various sectors of economy and as a result of degradation. Increasing pressure on these resources is accelerating the degradation processes. Developing strategies that promote conservation of resources, reclaim the already degraded resources and ensure the much needed productivity gains would be crucial in our future efforts to increase agricultural production to ensure individual and national food and nutritional security. On individual front, farmers owning salt affected lands have either no or very little earning from their precious resource. While introduction of irrigation raised hopes of higher productivity, these farmers lost within few years even what they were getting before irrigation. Reclamation of the degraded lands could bring these families back into the rural mainstream.

Table 1.18 Physico-chemical characteristics of a saline soil from Jind district (Haryana)

Depth (cm)	pH_s	EC_e (dS/m)	Ionic composition (mmol/L)							Boron (mg/L)
			Ca+Mg	Na	K	CO_3	HCO_3	Cl	SO_4	
0-11	7.5	4.5	20.5	35.6	0.6	Nil	1.0	35.5	21.5	11
11-30	7.6	4.6	30.7	22.6	0.5	Nil	1.0	40.5	10.0	6
30-56	7.8	4.8	32.6	23.6	0.3	Nil	1.0	37.5	12.7	6
56-76	7.8	5.6	30.5	27.3	0.2	Nil	1.0	39.5	14.5	7
76-141	8.0	8.3	41.5	42.8	0.2	Nil	6.0	58.0	22.5	10
141-200	8.0	12.7	50.7	74.1	0.2	Nil	10.0	70.5	39.5	14

Depth (cm)	Sand	Silt	Clay	$CaCO_3$ (%)		CEC cmol (p^+) kg	OM (%)	ESP
	(%)			<2 mm	>2mm			
0-11	69.3	21.2	9.5	Nil	-	4.4	0.5	6.8
11-30	61.5	24.9	13.6	0.8	-	5.7	0.3	7.0
30-56	60.9	21.8	17.3	2.9	-	7.8	0.3	7.6
56-76	55.7	21.8	22.5	4.4	7.0	8.9	0.2	8.9
76-141	57.0	21.6	21.4	8.0	35.0	8.9	0.2	8.9
141-200	56.1	22.0	21.9	16.8	15.0	7.6	0.1	9.2

When upscaled, the productivity loss at individual farms adversely affects the production and productivity of the irrigation commands. As a result projections made at the time of project planning are not realized. In the USA, the annual loss from the salt affected irrigated area has been estimated at 11.4 billion US \$ (1 US\$ = Rs. 54). The productivity loss in India in general is estimated at Rs. 10,000/ha. The data from the Chambal command however, put the figure at Rs. 21,000/ha. According to a conservative estimate made recently, loss of crop production due to water logging and salinity in the Tungabhadra project in Karnataka is assessed at about Rs. 250-300 million per year. Even a simple calculation on the anticipated losses in irrigation commands could be quite mind-boggling. For example if one ha of irrigated land goes out of cultivation, the investment of more than Rs. 150,000 is blocked. The interest on this investment alone is Rs. 15,000 per annum. Assuming a production of about 6 t/ha, a loss of about Rs. 50,000/ha would be entailed if the irrigated area remains out of cultivation. Assuming this area to be about 5.76 M ha, annual loss is estimated at Rs. 375,000 million. Thus, there is ample justification to spend about Rs. 375,000 million on prevention and reclamation of water logged salt affected lands.

Besides the economic losses, other social and health related issues calls for an action plan to reclaim water logged salt affected lands. Huge damage to infrastructure and flora and fauna has been experienced in few irrigation commands due to increased extent and degree of water logging and soil salinity. Rural housing is particularly vulnerable to damage as *kutcha* structures fall easily. Both human and animal population is prone to increased incidences of water borne diseases. Since, investments on health related programs can decidedly be brought down through reclamation; it is sufficient justification for investments in land reclamation programs.

As said before, land is an inelastic finite resource. Demand for the land is increasing as all the sectors are vying for this resource. While housing and industrial sectors could pay more for the same land, prime agricultural lands are being usurped by these sectors. As such, any kind of wastelands and more so the salt affected lands have to be reclaimed to compensate for the loss of agricultural lands to other sectors. Similarly, demand for agricultural products is increasing both due to increasing population and improving standard of living. With liberalization of economy, demand scenario has further changed. We need to generate increasing surpluses for conversion to value added products, which could be used for domestic use and exports. Apparently, both these objectives could be met by planning and investing on land reclamation programs.

2

DRAINAGE METHODS FOR RECLAMATION

Saline and alkali lands suffer from water logging either due to surface stagnation or shallow water table. While surface stagnation problem is common to saline and alkali lands, shallow water table of poor quality ground water is often associated with saline soils. In some situations, alkali lands could also suffer from shallow water table but often the ground water quality is either good or alkali in nature. From management point of view, waterlogged salt affected soils can be classified in the following categories:

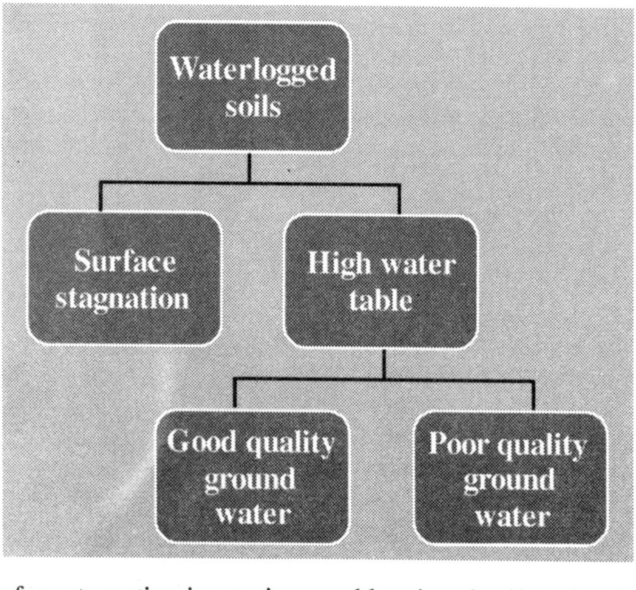

Surface stagnation is a serious problem in salt affected soils during monsoon season as rainwater cannot be absorbed in adequate quantities in alkali soils due to poor infiltration rate and in saline/alkali soils due to shallow water table. Since the formation of salt affected soils is due to inadequate drainage, the solution lies in improving drainage through artificially constructed systems. Both the surface and the subsurface drainage need to be augmented to minimize adverse effects of water logging. Improved surface drainage not only protects the crops from adverse effects of water stagnation, it also helps to reduce the cost of subsurface drainage. Amongst the various subsurface drainage techniques, horizontal subsurface drainage appears to be the most promising solution. It not only controls the water table but also facilitates salt leaching. Several other options have also been used to suit location specific situations. In this chapter, appropriate surface and subsurface drainage technologies for reclamation of waterlogged saline/alkali lands are discussed.

DRAINAGE NOMENCLATURE

Drainage has been classified into the following two main groups:

- Surface drainage
- Subsurface drainage

While in surface drainage, it is intended to remove water from the soil surface, in subsurface drainage, control of ground water is sought by removing water from the shallow water table. In salt affected soils, control of salinity is also sought using water as the transporting agent. Most common sub-classification of the subsurface drainage has been based on materials/structures used to drain the lands. On this basis subsurface drainage is classified as:

(a) Tile/Pipe drainage

(b) Tube well or vertical drainage

(c) Bio-drainage

As the names suggest, in the first case cement concrete tiles, plastic pipes or pipes of any other suitable materials are used to drain the lands. If tiles or pipes are not used then the pipe drainage is known as pipe less or mole drainage. In the second case, tube wells are used to drain the lands. In bio-drainage, as the name implies, a biological living material, an exotic tree, is used as the drain material. Other names for these systems are the horizontal drainage system and the vertical drainage system. While horizontal drainage system is used to designate the tile/pipe drainage system; tube well and bio-drainage respectively are designated as vertical drainage. The names are derived as the drain axis in the first case travels along the soil surface (horizontal) while in the second and third case it is perpendicular to the soil surface (vertical). Some other classification may also be prevalent but we restrict here to the nomenclature discussed in this section.

SURFACE DRAINAGE

Flat or nearly flat areas, heavy soils with low permeability and lands in humid or tropical and sub-tropical regions where high intensity storms are common, are subject to surface inundation and therefore, require surface drainage. Surface drainage problem may or may not be of local origin. In low lands it might occur due to the ingress of surface runoff from uplands. Overflow from rivers or natural channels sometimes contribute largely to the drainage problem of an area. The following factors contribute to the problem:

- Uneven land surface causing hindrance in the natural flow of runoff from the area
- Inadequate capacity of the drainage channels during critical periods
- Inadequate outlet conditions
- Back water flow

Surface drainage is carried out to meet one or more of the following objectives:

- In flat areas for timely removal of water so as to allow optimum growth of plants
- In sloping areas for orderly removal of water such that it does not cause erosion and
- For the collection and disposal of surface irrigation runoff.

Surface drainage has often been considered as a panacea to all drainage problems. However, it is now known that it cannot help to lower the water table effectively although rate of build-up of the water table could be reduced through efficient surface drainage. In the long-term, surface drainage reduces installation and operational costs of subsurface drainage systems.

Surface Drainage Requirement and Crops

The adverse effects of temporary water stagnation on the land surface depends on plant species, the variety of the crop, physiological stage of growth at which water stagnation occurs (crop may be more sensitive at one stage compared to another), the time and duration of water logging, the light intensity, temperature and the fertility of the soil. Injury to crops is particularly severe when the crops are flooded on hot days. The damaging effects are generally more when water logging conditions are accompanied by salinity/sodicity. It may be noted that no single factor affects the crop yield and generally, it is a combination of poor aeration, reduced uptake of plant nutrients, imbalance between the uptake of various ions and toxicity effects of various elements/cations.

Not all crops suffer equally from inundation or root submergence. The damage which is dependent upon the degree and duration of submergence varies from one crop to another (Table 2.1). In a semi-reclaimed alkali land, one day water stagnation reduced the yield of *berseem* (seed) by only 2% whereas wheat and mustard yields reduced by 8% (Gupta *et al.*, 2004). The yield loss increased and varied from 20% in sorghum to 48% in *berseem* (seed) when water stagnation occurred for six days. For the same duration, yield loss in wheat increased to 39%. Grasses and forages are usually more tolerant to inundation as compared to cereals or other crops. In the Indian context, only limited data for few crops are available in this regard. With the available information to date, few general observations could be made:

Table 2.1 Yield in drained plots and percent yield reduction due to water stagnation of various durations for tested crops

Crop	Yield (t/ha)	Yield reduction# over drained (%) for water stagnation (days)			
	Drained	1	2	4	6
Sorghum	4.13	3	11	16	20
Pearl millet	2.22	6	15	22	27
Pigeon pea	1.52	4	14	18	21
Wheat	4.20	8	17	27	39
Barley	3.65	4	7	13	25
Mustard	1.43	8	16	22	29
Berseem (seed)	0.48	2	21	35	48
Sunflower	1.86	13	19	26	30

Yield reduction due to water stagnation is relative to drained plots

- The crops, in general, are more sensitive to water stagnation at early than later growth stages

- The yield reduction increases with increasing degree and duration of submergence.

General guidelines for the periods within which stagnated water should be removed from crop lands are given in Table 2.2.

Table 2.2 Permissible water stagnation period for different crops

Crop	Permissible water stagnation period
Vegetables	1
Oil seeds	2
Cotton, maize	3
Rice, sugarcane	7

Field Drainage Coefficient

Drainage coefficient is defined as the amount of water that runs off from a given area and is to be removed within 24 h. Commonly, the unit is depth of water per day (mm/d). Many a times, the drainage coefficient is also expressed in flow rate (m^3/s) per unit area. It is not often possible to determine drainage coefficient with any degree of accuracy although it is the key factor in establishing the needed capacity of the drainage system. Several methods are used to evaluate the drainage coefficient. The use of a particular method depends mainly on the availability of required data. The simplest method to assess surface drainage coefficient for farm drainage consists of the following steps (Finkel, 1983):

- Evaluate maximum 24 h rainfall depth which may occur with a probability of 10-20%. Frequency analysis procedures are commonly employed for this purpose
- Evaluate the basic infiltration rate of the soil and estimate the potential evapotranspiration
- Estimate the tolerance of crop to surface inundation. For example let us say it is 'n' days. Multiply the infiltration rate of the soil (mm/d) and expected potential evapotranspiration (mm/d) by the number of days, 'n'
- Subtract the value calculated in preceding step from the value of maximum 24 h rainfall. The resulting value divided by number of days, n will give the drainage coefficient in depth of water to be removed in 24 h (mm/d).

Mathematically, these steps can be put in the form of an equation as below:

$$q = \frac{1}{n}[R - n(E + 1)] \tag{2.1}$$

Here q is the drainage rate in mm/d, R is the rainfall in mm, E and I are potential evapotranspiration and infiltration rate in mm/d and n is number of days. When n is greater than one, it may be useful to increase the duration of rainfall from 1-day maximum to n-days maximum (R_n). With this change, drainage coefficient will be nearer to reality. Similarly, for rice crop, where some depth of water is allowed to stand in the fields after the drainage period, necessary correction in eq. (2.1) could be made by subtracting this depth, D_r, while calculating the drainage coefficient for this crop. With these changes, eq. (2.1) can be written as:

$$q = \frac{1}{n}[R_n - n(E + 1) - D_r] \tag{2.2}$$

A number of workers have estimated drainage coefficient for different agro-climatic regions of the country. Most comprehensive analysis made by Gupta *et al.* (1971) revealed that drainage coefficients in India vary from a low value of 2 mm/d to about 40 mm/d depending upon the crop (rice, grain or vegetable) and the climatic conditions.

Drainage Components and Field Practices

A surface drainage system essentially consists of the following three parts:

- Collection system
- Disposal system
- Outlet

Collection system mainly concerns on farm drainage with disposal and outlets forming a part of the overall system. A major component of the disposal system and the outlets are constructed and maintained by the State Governments. Some of the commonly used formulae to design main and collector drains in various states are listed in Table 2.3. An example to illustrate the design of a collector is given in Table 2.4. Common cross sections for various components of surface drainage adopted in the Chambal command of Rajasthan are given in Table 2.5.

Table 2.3. State wise formulae or guidelines used in design of main/ collector drain

State	Formulae / guideline	Remarks
Punjab and Haryana	$Q= C^{*}\sqrt{A}/5$ (old)	
	$Q= C\sqrt{A}/6$	Up to 260 km^2
	$Q= C\sqrt{A}/4$	261-650 km^2
	$Q= C\sqrt{A}/2$	>651 km^2
Uttar Pradesh and Andhra Pradesh	R=0.11 A	
Madhya Pradesh	$Q= 0.22^{\#} A$	For A>79 km^2
	Q= 0.44 A	For A<13 km^2
Tamil Nadu	Q= 0.44 A	A= 13 km^2
	$Q= 0.6A^{2/3}$	A> 52 km^2
Odisha	$Q= F A^{5/6}$	F= 0.3 for discharge of 10-12 cusecs

Q = discharge in cumecs, A = catchment area in km^2 and *C is a coefficient, which depends upon rainfall. Values vary from 3.5 for 500 mm rainfall to 35 for rainfall in the range of 750- 1000mm. # The values vary from 0.22 to 0.44 for areas in between 13 to 79 km^2. 1 cumecs = 35.3 cusecs

Table 2.4. A typical design example for a collector drain

1	**Salient information**	
	Catchment area	100 ha
	Land slope	0.10%
	Soil type	Sandy loam
	Average annual rainfall	1220 mm
2	**Assumed parameters**	
	Side slope	H: V = 1:1 (Depends upon soil type)
	Channel bed slope S=0.10%	0.001 (As per survey)
	Run off coefficient	F=0.30
3	**Design run off**	$Q = F \times A^{5/6}$ A = Catchment area in km^2 $Q = 0.30 \times (1.0)^{5/6}$ = 0.30 cumecs or 9.96 cusecs
4	**Design of section**	
	For efficient channel section	b=2d (Assumed); R = 0.6213 d

	Where
	R = hydraulic radius
	d = depth of flow
	$A = 3d^2$
Velocity of flow	$V = 1/n\ (R^{2/3} \times S^{1/2})$; For n = 0.035
	$V = 1/0.035 \times (0.6213d)^{2/3} \times (0.001)^{1/2}$
	$= 0.6578\ d$
	$Q = A \times V = 3\ d^2 \times 0.6578\ d^{2/3}$
	$= 1.9736 \times d^{8/3}$
Solving	$Q = A \times V$
	$0.30 = 1.9736\ xd^{8/3}$
	d = 0.0.49 say 0.50 m
With 20% free board total depth of flow	D = 1.20 x 0.50 = 0.60 m
	b= 2d = 1.20 m
	T= b+2 x z x d (T=Top width)
	z = side slope = 1
	T = 2.40 m
Therefore, the design section is	
Top width, bottom width and depth	2.4 m, 1.20 m, 0.6 m

Table 2.5 Cross section of surface drains adopted in Chambal Command, Rajasthan, India (design runoff rate: 1.1 L/s/ha)

Type of drain	Base width (m)	Side slope (H:V)	Depth (m)	Density (m/ha)
Main drain	0.75-1.20	1.5: 1	2.0	7 -10
Seepage drain	0.60-0.75	1 :1	1.0	15 -20
Carrier drain	1.0-1.20	1:1	1.0-1.20	15-20
Field drain	0.60-0.75	1:1	0.9-1.20	20-40

Land Grading/Leveling

Land grading/land leveling is synonymously used in surface irrigation and drainage. Land grading is more descriptive than land leveling where a sloping plane surface than a level surface is designed. Land grading is a onetime operation in which pre-decided design slope is achieved. The design slope should meet the requirements of both the surface irrigation and drainage. For drainage, the grade in the direction of drains can be in the range of 0.05 to 0.5% but a preferable slope is 0.1%. Currently, laser land leveling is becoming quite popular. Results reveal about 20% saving in water and 10-15% increase in the yield of wheat with laser as compared to conventional leveling. The increase in yield and water saving is attributed to improved land leveling index showing a perceptible improvement (0.8-0.9 over 2.1) with laser than conventional leveling (Table 2.6). The laser levelers are now available under custom-hiring mode obviating the need for investment by individual farmers.

Table 2.6. Performance parameters of laser land leveling

Field	Mean value	Range		LI	
		Before	After	Before	After
1 Laser	2.102	2.055-2.155	2.074-2.124	1.56	0.80
2 Laser	2.860	2.790-2.940	2.830-2.890	2.29	0.90
3 Manual	2.860	2.784-2.980	2.817-2.917	3.00	2.10

Drainage Systems

A number of surface drainage systems are in vogue. Some of them are bedding system, parallel field drain system, the random drain system and the parallel open ditch system.

Bedding system

Bedding system is probably the oldest of the surface drainage systems. It is suitable for soils with poor hydraulic conductivity with slopes not exceeding 1.5%. The level surface of the field is converted into a sloping area by plowing several times in such a way so as to always throw the back furrow in the channel. The beds are always kept parallel to the greatest slope of the field such that the dead furrow serves as a field drain. The bed width depends upon several factors including the slope, the land use, soil permeability, farm machinery to be used and the nature of the farming operations. Based on hydraulic conductivity, K, the bed width varies from 8 to 30 m (Table 2.7).

The length of the beds could be in the range of 100 to 300 m. The height of the crown from the bottom of the furrow should not exceed 45 cm for pastures and 20 cm for arable lands (Fig. 2.1). Field laterals are provided perpendicular to the field drains which collects the water from these drains to carry it to the main drain. The bedding system has many disadvantages particularly for row crops. The flow to the furrows is obstructed when rows are parallel to the dead furrows. Moreover, the slope of the furrow may not always be sufficient to take care of the overland flow. Since the system moves the top soil from the side of the beds to the middle, fertility gradients are established at least in the initial years. Another limitation of this method could be due to restrictions in mechanized farming. Maintenance cost is also high as the beds would require regular maintenance. The method is most useful for pasture lands.

Table 2.7 Bed widths in relation to hydraulic conductivity of soils

Hydraulic conductivity (m/d)	Bed width (m)
K≤ 0.05	8 - 12
0.05< K≤ 0.10	15 - 17
0.10< K≤ 0.2	20- 30

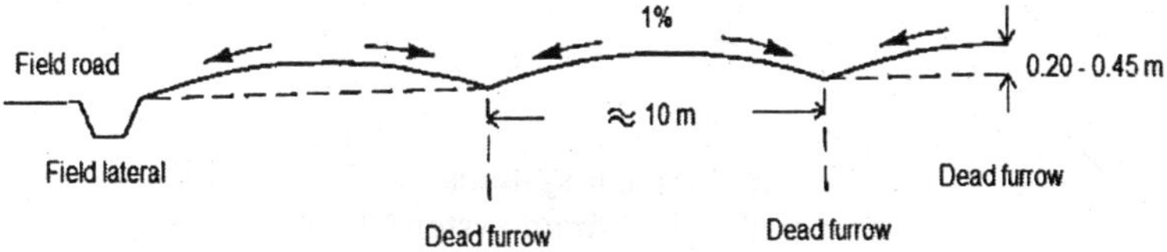

Fig. 2.1 Cross-section of a bedding system of land drainage

Local adaptations of the system are emerging in many countries particularly under heavy textured clay soils. One such alternative known as camber beds is being researched in Ethiopia. Camber beds consist of 6 m wide beds followed by 1 m furrow (Fig. 2.2). Camber beds are constructed using a mould board plow where the soil is turned from the furrow side to form the beds. While the crop is planted on the 6 m wide beds, furrow serves to drain the water. Yields about 50% higher than the conventional farming have been reported.

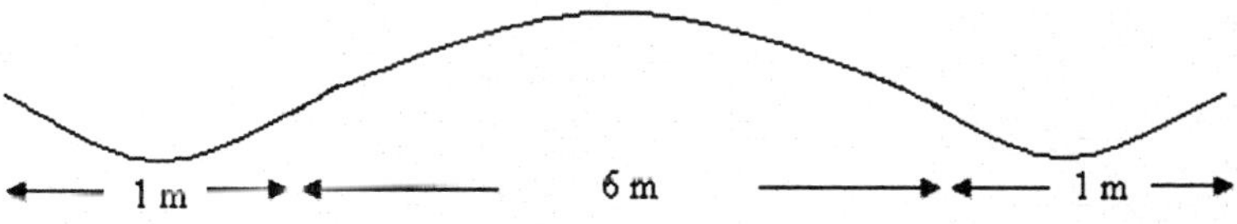

Fig. 2.2 A schematic view of a camber bed

Parallel field drain system

The parallel system is used when the surface is uniform and has few noticeable ridges or depressions. The channels in this case are spaced farther apart than a bedding system and have greater capacity than the dead furrows. Although the design and layout is similar to bedding system, it is not necessary that drains are spaced equally or they drain on both sides. The depth of the drain should be minimum 20 cm having a minimum cross section area of 0.45 m^2. The side slopes are 8:1 or flatter. The row or field drains are perpendicular to the field lateral. The system is most suitable for fields where mechanization is practiced. Land smoothening of field surface between laterals is necessary so that land remains well graded.

Random system

This system is best suited to drain scattered depressions. Field or row drains and field laterals meander from one depression to another to drain the same to the outlet ditches. Location of the lateral ditches is governed by the topography of the land. The system is most suitable where depressions are deep and it is technically or economically infeasible to fill up the depressions. The slope of the channel can be as flat as 10:1 when drain depth is more than 60 cm or 4:1 if fields are farmed parallel to the drains. The recommended minimum cross section area is 0.45 m^2. The channels follow a route that requires minimum cut and ensures minimum interference in the farming operations. The system costs much less and provides more returns on investments. The maintenance cost is relatively more than other systems of drainage.

Parallel lateral open ditch system

The parallel lateral open ditch systems besides providing surface drainage also help to control the ground water table. The layout of the system is similar to the field drain system but ditches are deeper. When financial or other constraints do not allow construction of tile/pipe drainage, this system of drainage is recommended till finances are generated or made available. The system is also referred to as open subsurface drainage system

Government Support

Ministry of Water Resources, GOI restructured the CAD program as CAD&WM during the year 2004. It supports surface drainage and land reclamation activities. The revised ceiling and funding pattern for relevant components is shown in Table 2.8.

Table 2.8 Revised guidelines on expenditure and sharing of expenditure by various stakeholders

S. N.	Items	Cost norm (Rs./ha)	Central share	State share	Farmer's share
1	Construction of field channels	Rs. 10,000/-	50%	40%	10%
2	Construction of field drains	Rs. 4,000/-	50%	50%	
3	Reclamation of waterlogged area	Rs. 15,000/-	50%	40%	10%
7	Renovation and desilting of existing irrigation tanks within the irrigated command	Rs. 15,000/-	50%	40%	10%

Hydraulic Design of Field Drains

Hydraulic design refers to working out the hydraulic dimensions of different components of the cross sections of the drains to carry the designed runoff. The most commonly adopted approach is the use of the Manning's uniform flow formula to estimate the capacity of the drain:

$$V = (1/n)\,(R^{2/3})\,(s^{1/2}) \tag{2.3}$$

$$Q = A\,V \tag{2.4}$$

Here, V is the mean velocity of flow in the channel, m/s, n is the Manning's roughness coefficient, R is the hydraulic radius of flow, = A/P, m, A is the flow area of channel, m^2, P is the wetted perimeter, m, and s is the channel bed slope, m/m and Q is the design discharge, m^3/sec.

Manning's roughness coefficient, n, is selected from the standard values for a variety of channel flow conditions and s is usually decided on the basis of local land slope. The average indicative values of n for excavated unlined drainage channel are in the range of 0.02 to 0.06 (Table 2.9). Bed slopes may not be less than 0.2% to avoid siltation and more than 0.5% to avoid channel scouring. Maximum permissible velocities for various situations are given in Table 2.10 and Table 2.11. Table 2.10 also lists the maximum permissible side slopes. Side slopes generally become more stable once they have become vegetated but design should mostly be based on the initial period after excavation.

Table 2.9 Values of Manning's n for wet cross section well covered with weedy or grassy vegetation

Channel description	n-value
(i) Small channel (water depth <0.75 m)	
Sandy soil	0.050
Clayey soil	0.060
(ii) Medium channel (water depth 0.75-1.5m)	
Sandy soil	0.035
Clayey soil	0.050
(iii) Large channels (water depth >1.5 m)	0.020-0.025
(iv) Concrete lined channel	0.012-0.017

Source: Smedema and Rycroft (1983)

Table 2.10 Limitations on flow velocity and side slope in drainage channels

Soil type	Maximum permissible mean flow velocity (m/s)	Maximum permissible side slope (vertical : horizontal)
Fine sand	0.15-0.30	1 : 2 to 3
Coarse sand	0.20-0.50	1 : 1.5 to 3
Loam	0.30-0.60	1 : 1.5 to 2
Heavy clay	0.60- 0.80	1 : 1 to 2

Table 2.11 Permissible velocities in non-vegetated channels

Material excavated	Clean water	Velocity (m/s)	
		Water transporting colloidal silt	Water transporting sand, gravel or fragments
Fine sand, colloidal	0.45	0.75	0.45
Sandy loam, non-colloidal	0.53	0.75	0.80
Silt loam, non-colloidal	0.60	0.90	0.60
Alluvial silt, non-colloidal	0.60	1.05	0.60
Ordinary firm loam	0.75	1.05	0.68
Stiff clay, very colloidal	1.13	1.50	0.90
Fine gravel	0.75	1.50	1.13
Coarse gravel, non-colloidal	1.20	1.80	1.95
Cobbles and shingles	1.80	1.80	1.50

Source: Fortier and Scobey (1926)

Channel cross section

Channels with a semicircular wet cross section are hydraulically the most efficient (these require minimum excavation per unit wet cross section). However, drainage channels with a trapezoidal cross section have been found to be most suitable and workable. Following factors are taken into account while designing the cross section of channel:

- Drain depth would depend upon whether it is to carry only the surface runoff or is to control the ground water table also
- Bed gradient depends upon outfall conditions
- Bed width also depends on the peak design discharge at different points
- Design velocity should be selected so as to maintain the ditch cross section.

The bottom width for the most efficient cross section and minimum volume of excavation for a trapezoidal channel is given by

$$b = 2d \tan \theta/2$$

Where θ is angle formed by side slope. It may not always be possible to design for the most efficient cross section. As a compromise between hydraulic efficiency and other considerations, the ratios between bed width (b), and water depth (d) should approximately be adhered to (Table 2.12; Smedema and Rycroft, 1983). Once b and d are decided, cross sectional area and wetted perimeter of the channel are worked out be using the relations:

$$\text{Cross sectional area } A = bd + zd^2$$

$$\text{Wetted perimeter } P = b + 2d\,(z^2 + 1)^2$$

Where, z is the side slope z : 1 (Horizontal : Vertical)

Table 2.12 Recommended bed width and depth ratios for open channels

Drain depth	**Recommended b/d ratios for trapezoidal channels**
Small channel, d=0.75 m	1 for clay to 2 for sand
Medium channel, d=0.75 to 1.5 m	2 to 3
Large channel, d=1.5m	3 to 4

Depending on the method of construction, minimum sections should have:

- Bed widths of 0.50 -1.0 m

- Side slope of 1:1 or 1: 1.5, and
- Water depth of 0.30 -0.50 m.

Freeboard is meant to provide a margin of safety in case the actual discharge exceeds the designed discharge capacity of the drain. Freeboard is less when the design discharge associated with the higher return period is more. Typical guidelines for freeboard are:

- 30 cm for channels having capacity < 2 cumecs
- 50 cm for large channels having capacity > 5 cumecs.
- For small channels having 0.25-0.5 cumecs discharge, the freeboard ranges from 10 to 25% of the designed depth.

Berm is provided for stability of slope. It depends on the depth of the drain and type of soil and adjacent land use. Minimum recommended berm widths are as below:

Depth of drain (m)	Minimum berm width (m)
0.6 -1.2	1.2
1.2 - 1.8	1.8
1.8 – 2.4	3.0

Considering all these issues, a typical cross-section of a drainage channel would appears as shown in Fig. 2.3.

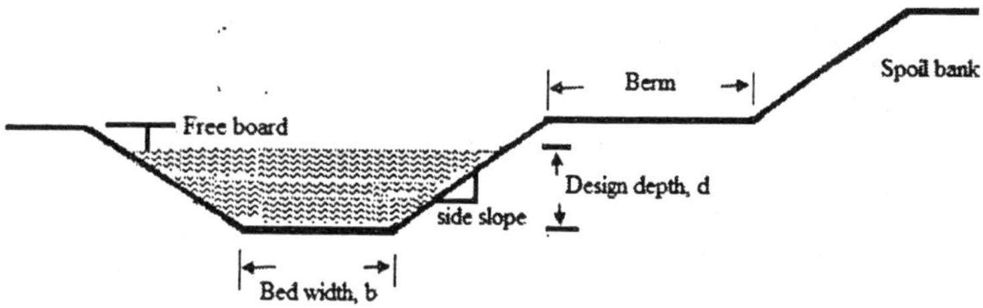

Fig. 2.3 Cross-section of a drainage channel

MAINTENANCE: OPEN DRAINS

Poor design, improper construction and lack of timely and adequate maintenance could be responsible for the failure of an open ditch system. Even a well designed and installed system would fail to function well if it is not maintained properly. After initial construction, open drains may lose their effectiveness due to one or more of the following reasons.

- Sedimentation

- Vegetation growth
- Bank erosion
- Damage to side slopes

Maintenance has been divided into two categories i.e. preventive maintenance before failure and corrective maintenance after partial or complete failure. Preventive maintenance is usually cheaper and prevents or delays the failure of the system. Likewise timely maintenance is more economical than delayed maintenance.

Preventive Maintenance

Prevention of sedimentation, control of excessive vegetative growth in the channels, prevention of bank erosion and repair of damaged portion of the drain are some of the works that are needed to prevent drain failure. Sedimentation occurs when the water entering the drain carries sediments. Minimizing the erosion in the watershed through the construction of simple drop structures, sedimentation basins and diversion ditches could reduce the sediment deposits in the drains. Scouring and deposition of silt may also occur along the bed of the drain particularly at points where the grade changes sharply and there is change in flow velocities. Settling basins and low check dams can be constructed under these situations. These activities would considerably reduce the clean out works in the drains.

Weed growth or vegetation reduces the capacity of the drains. As such, drains fail to carry the design discharge. Excessive weed growth in channels has been found to reduce the capacity of the channels by 75% or more. Control of excessive tree, brush, grass and weed growth along the sides of the channels as well as in the channels is an important preventive maintenance measure. Manual removal of weeds is cumbersome and expensive. Therefore, mechanical and chemical means should be employed to control weed growth. Mowing, grubbing, clearing, grazing, burning and spraying are used to control weed growth. Burning of weeds or spraying of weedicides may be required more than once in the growing season. The practice of grazing should be used with caution as it might damage the side slopes.

In a properly designed and constructed drain, the damage duc to bank erosion and side slopes will be minimal. Wherever necessary, the spoil banks should be seeded to grow permanent vegetation. Cattle and human trespass should be minimized to maintain side slopes.

Corrective Maintenance

After the first construction, changes in cross section, grade or alignment of the drain may be required for the drain to function well. Due to variation in soil type, side slopes at some places may cave in and fill the drain. The side slope at these vulnerable points needs to be reshaped to a more stable slope. Sharp curves causing meandering of the drains may necessitate modifications in the ditch alignment. Either the drain may be straightened or the curve may be flattened.

TILE/ PIPE OR HORIZONTAL SUBSURFACE DRAINAGE

In order to develop and sustain high productivity agriculture, agricultural lands in the humid and sub-humid regions and most irrigated lands need subsurface drainage. The challenge is to correctly evaluate the problem, design and install most economical drainage system, to maintain the system so as to ensure maximum benefits from the investments made.

A horizontal drainage system consists of the laterals, the collectors and the main drains. Lateral drains (Fig. 2.4) are laid parallel to each

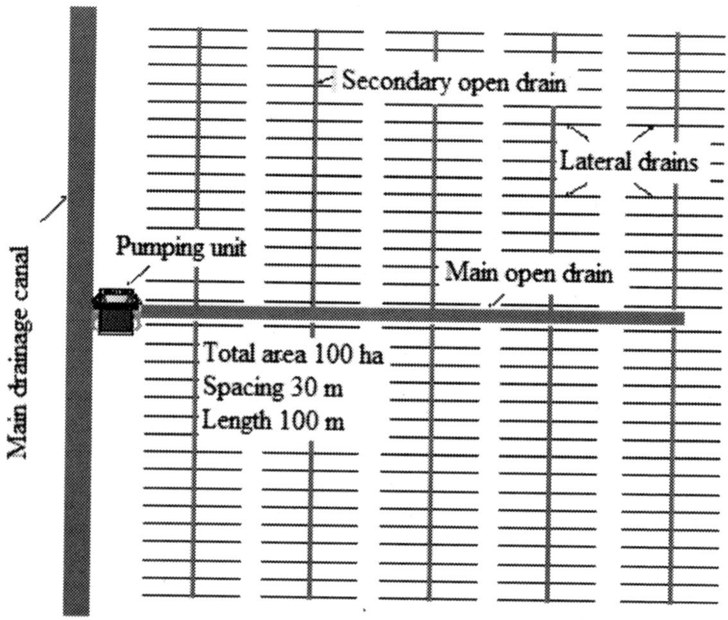

Fig. 2.4 A schematic view of the tile/pipe drainage system

other to cover the entire field. These drains accept the excess ground water from the soil above the drain level and primarily control the ground water table. The water entering into the laterals flows into the collectors which carry it to the main drain. The main drains carry the water to the outlet from where it is disposed out of the area either by gravity or by pumping from a sump (Smedema and Rycroft, 1983). The processes of salt and water flow are schematically shown in Fig. 2.5.

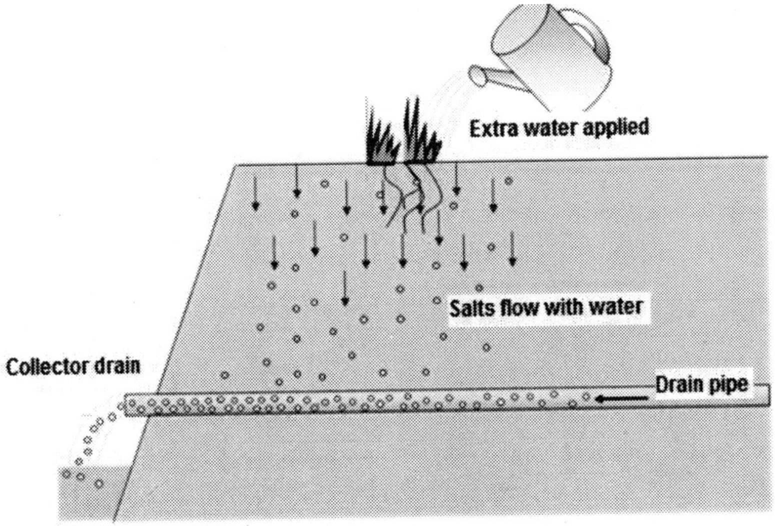

Fig. 2.5 The diagram showing processes of water table control and salt leaching in drained lands

Drainage Investigations

In order to determine the feasibility of the drainage and to design and install an efficient and economical drainage system, investigations of the site conditions is essential. Surveys and investigations for drainage projects should be carried out to reveal the following information

- Inventory of the project site conditions
- Causes of the excess water
- Identification of appropriate solution
- Data needed to establish drainage system design parameters

Following information would be required for design purposes:

- Topographic map

- Data on soil salinity/alkalinity, soil profile description and depth to impermeable layer etc.
- Position and fluctuations of the ground water table, depth to water table, artisan pressure and ground water quality
- Hydraulic conductivity/drainable pore space
- Crops and their tolerance to water logging
- Irrigation practices and water requirement

The information is obtained by systematic surveys and investigations. The amount of investigations required would mainly depend upon the investigator's knowledge of the area, manpower and the funds at disposal, amount of data already in hand or available with different agencies operating in the area, size of the project and data available for similar projects executed earlier. Projects covering large areas would require more intensive investigations.

Water Transmission Characteristics

The two most important soil characteristics that are commonly used and referred to in drainage investigation and design are: (i) the hydraulic conductivity and (ii) the drainable pore space. Reliable estimates of these soil characteristics are the backbone of a horizontal subsurface drainage system.

There are several methods to determine hydraulic conductivity. For drainage design, *in-situ* determination of hydraulic conductivity is desirable. While auger hole method is used to determine hydraulic conductivity of soil layers below the water table, the inverse auger hole method is employed to determine the hydraulic conductivity of soil layers above the water table. The auger hole method is applicable to both homogeneous and layered soils provided the pertinent layers are 10 times or more thicker than the radius of the auger hole. The method is ideal for depths up to 5 m. Piezometer method is used to determine the hydraulic conductivity of soil layers as thin as 10 cm and for depths up to 8 m. Readers may refer a standard text book on drainage for more information on conducting *in- situ* hydraulic conductivity tests (Ritzema, 1994).

Besides laboratory methods are also used to determine soil hydraulic conductivity but they have little applicability in drainage design. A first guess of the hydraulic conductivity can also be made on the basis of soil texture (Table 2.13).

Table 2.13 Relationship between hydraulic conductivity and soil texture

Soil texture	Hydraulic conductivity (cm/h)
Clay	.0025 - .0380
Silty clay	.0381 - .0635
Silty clay loam	.0635 - .1016
Clay loam	.1016 - .1778
Sandy clay loam	.1778 - .3810
Silt loam	.3810 - .8890
Loam	.8890 -1.7780
Very fine sandy loam	1.778 - 3.048
Fine sandy loam	3.048 - 4.318
Sandy loam	4.318 - 6.350
Loamy fine sand	6.350 - 8.890
Loamy sand	8.890 -11.430
Fine sand	11.430 -15.240
Sand	15.240 -20.320
Coarse sand	20.320 -30.480

Drainable porosity

The drainable porosity also known as effective porosity and specific yield is defined as the volume of water released or taken into storage in an unconfined aquifer per unit cross sectional area per unit decline or rise in water table as a fraction of total volume of the soil. For drainage conditions, drainable porosity of a soil relates to the amount of fluctuation of the water table to the amount of ground water added or drained from the system. To determine the build-up of the water table from each increment of recharge, the depth of each recharge should be divided by the drainable porosity.

Both laboratory and field determinations of drainable porosity are expensive and time consuming. A large number of tests are needed to obtain a representative drainable porosity of the area. On the basis of considerable data, a general relationship (Fig. 2.6) has been developed between the hydraulic conductivity and drainable porosity (US Bureau of Reclamation, 1971). Since the fluctuation of the water table in a drained area takes place in the zone between the drains and the maximum allowable water table height, the drainable porosity corresponding to the weighted average hydraulic conductivity of this zone should be used in drain spacing calculations in most cases.

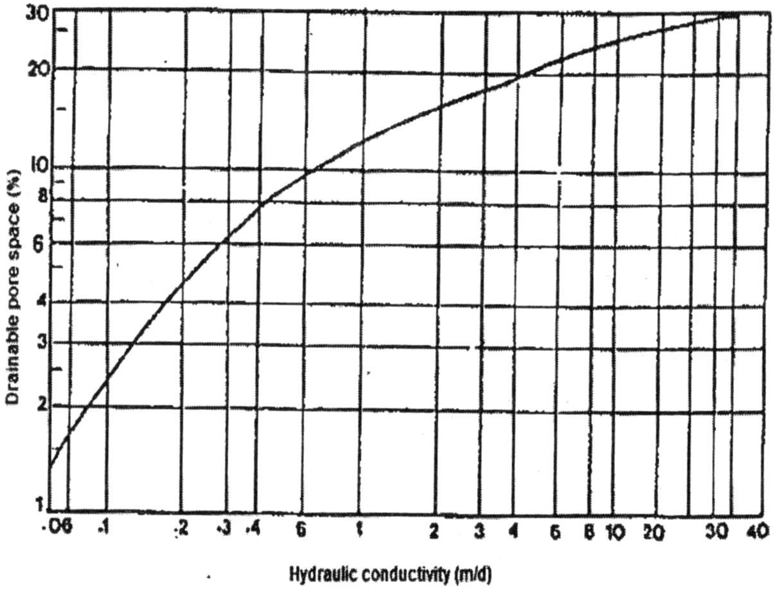

Fig. 2.6 Relationship between hydraulic conductivity and drainable pore space

Another approach to estimate drainable pore space is on the basis of soil texture and soil structure which have considerable influence on the drainable pore space. The representative volumetric drainable porosity values for different textures of soil can be taken as 3-5% for heavy clays, 6-9% for medium textured soils and 10-15% for sandy soils. Exact estimates can be made from the data presented in Table 2.14. Whenever possible, the drainable porosity should be determined from the measurements of drain discharge and draw-down of existing or pilot drains.

Drainage Coefficient

The drainage coefficient controls the drain spacing, the size of the field laterals and the capacity of the collectors and pumping station (if any). As such, it needs to be assessed quite accurately. Although elaborate procedures are in vogue to evaluate drainage coefficients, the range of drainage coefficients assessed/observed at few places in the country is reported in Table 2.15. Based on further analysis, Gupta (2011) prepared guidelines according to which drainage coefficients range from 1-5 mm/d (Table 2.16).

Table 2.14 Influence of soil texture and structure on drainable pore space

Texture	Structure	Drainable pore space
Clay Heavy clay loam	Massive, very fine or fine columnar	1-2
Clay Clay loam Silty clay Sandy clay loam	Very fine or fine prismatic, angular blocky or platy	1-3
Clay Silty clay Sandy clay Silty clay loam Clay loam Silty loam Silt Sand clay loam	Fine and medium prismatic, angular blocky and platy	3-8
Light clay loam Silt Silt loam Very fine sandy loam Loam	Medium prismatic and sand –angular blocky	6-12
Fine sandy loam Sandy loam	Coarse sub-angular blocky and granular, fine crumb	12-18
Loamy sand Fine sand	Medium crumb and single grain	15-22
Medium sand	Single grain	22-26
Coarse sand Gravel	Single grain	26-35

Source: United States Bureau of Reclamation, 1971

Table 2.15 Subsurface drainage coefficients from different locations (observed from test plots)

Site	Rainfall (mm)	Rate (mm/d)	Recommended range (mm/d)
Chambal (Rajasthan)	850	3.0	2.5-3.5
Sampla (Haryana)	600	2.5	2.0-3.0
Hisar (Haryana)	400	2.0	1.5-2.5
Dabhou (Gujarat)	800	4.0	3.0-5.0
Mundlana (Haryana)	500	5.0	5.0-7.0
Kailanakhas (Haryana)	500	6.8	5.0-7.0
Muraj (Gujarat)	500	2.8	2.0-4.0

* Variation in the range of 600-1400 mm

Table 2.16 Guidelines on subsurface drainage coefficients

Climatic conditions	Range (mm/d)	Optimum value (mm/d)
Arid	1-2	1
Semiarid	1-3	2
Sub-humid	2-5	3

Source: Gupta (2011)

DESIGN OF SUBSURFACE DRAINAGE SYSTEM

To design a subsurface drainage system the following main items have to be considered.

- Laterals: Depth and spacing of field laterals; diameter and slope for carrying capacity
- Collector: Diameter and slope for carrying capacity
- Sump/outlet

Laterals: Depth of Drains

Some important considerations to determine drains depths are:

- The desirable depth to water table for proper aeration in the root zone
- The depth of drain placement such that the cost/ha is minimum (Fig. 2.7)
- Soil conditions such as unstable or hard soils and quick sand conditions
- The available depth of outlet drains, topography and minimum pipe slopes

For irrigated annual crops in loam, sandy loam and silt loam soils, average water table depth of 0.60 to 0.80 m is desirable. For clay loam and clay soils, the depth should be 0.8 to 1.0 m. Perennial crops and especially trees need deeper depths to water table. An average seasonal water table depth of about 1.0 m will provide reasonably favorable conditions in the root zone for most crops. The recommended depths to water table for various crops are listed in Table 2.17.

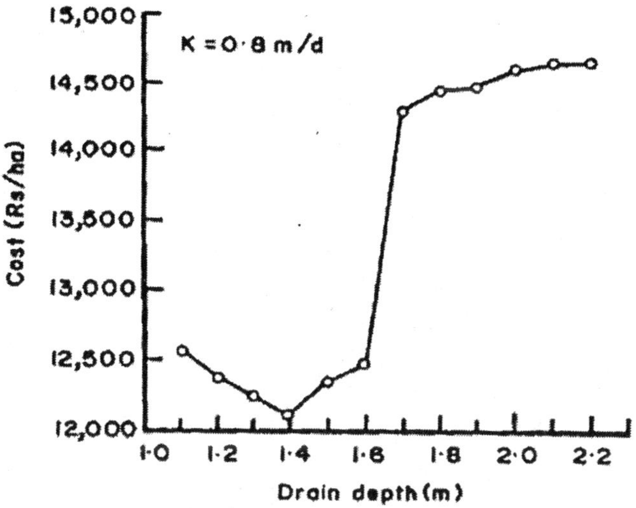

Fig. 2.7 Drain depth versus cost of drainage

Table 2.17 Optimum depth to water table for different crops

Crops	Depth to water table (cm)
Pearl millet	125
Wheat	30-50
Pea	50
Maize (Corn)	60-100
Soybean	125-150
Cow pea	75
Sorghum	75
Barley	100
Mustard	125
Gram (Chick pea)	100
Sugarcane	75
Safflower	100
Groundnut (summer)	100
Groundnut (winter)	50
Rice	<30
Green gram	60

Shallow versus deep drains

The deep drains allow wider drain spacing and provide larger open water and soil storage. In open drains, land lost with deeper drains is less than relatively shallower drains. Contrarily, a deep drain system will cost more due to higher excavation costs and the additional cost for larger pipes in the case of covered drains. An analysis made at CSSRI, Karnal showed that the cost is significantly affected by the drain depth. Initially the cost reduces with increasing depth but later it increases with increasing depth. A drain depth of 1.4 m and its corresponding drain spacing resulted in the lowest total cost for a sandy loam soils (Fig. 2.7). Sometimes, maintenance cost could also be a limiting factor in deciding the drain depth. In pumped outlets, extra cost of pumping due to larger elevation against which a pump operates must also be taken into account.

In practice drain depth varies greatly. The minimum and maximum values are commonly 0.5 m and 2 m respectively. Within this range, the depth of the drains should be selected in a manner such that the water table at midway between the two lateral drains is at a satisfactory depth at critical times. If a crop grows well at a depth to water table of 1 m then the water table at midpoint between the drains should be at least 1 m deep during the cropping season. The selected depth should also meet the minimum recommended soil cover criterion to avoid damage due to use of heavy machinery (Table 2.18). From practical view point, a depth of 1.0 to 1.8 m is optimum for salinity control (Table 2.19).

Table 2.18 Minimum recommended soil cover over tiles

Soil type	Depth of soil cover (cm)
Mineral	60
Deep peat and Muck	120
Organic	75

Table 2.19 Guidelines on drain depth

Outlet condition	Depth of the drains	Optimum depth (m)
Gravity	0.9-1.2	1.1
Pumped	1.2-1.8	1.5

Soil salinization and drain depth

There are two schools of opinion on this subject. While one advocates shallow drains, the other favors deep drains. The advocates of deep drains usually rely on the concept of critical depth to water table arguing that deep drains help to avoid secondary salinization besides other advantages which might accrue with deep drains. For example, deeper drains will allow larger lateral drain spacing. On the other hand, recent studies have conclusively proved that secondary salinization can be avoided provided flux at the water table could be kept downward. In such a case, shallow drains would be as effective as deeper drains. In monsoon climatic regions, when the shallow water table conditions prevail, the flux on an average is downward. Therefore, in the Indian context, shallow drains (depth 1.0- 1.8 m) would be quite effective in reclamation and management of saline soils.

Spacing of Lateral Drains

The water table between two parallel drains is usually curved, its elevation from the center of the drains being highest at midway between the drains. The factors that influence height of the water table in drained lands are:

- precipitation and other sources of recharge
- evaporation and other sources of discharge
- soil properties
- depth and spacing of the drains
- cross-sectional area of the drains
- water level in the drains

Several equations have been developed interrelating the above factors. These equations are either based on steady state or non-steady state condition of the water table. In the following sections one most commonly used equation each for the two situations is described.

Steady state

The steady state formulae have been derived assuming that the recharge intensity equals the drain discharge rate. As such, the water table is considered to remain at a constant position (Fig. 2.8). The most commonly used steady state equation to determine drain spacing is the Hooghoudt's equation, which for a two layered profile can be written as (Hooghoudt, 1940):

$$q = \frac{8K_b\,DH + K_a\,H^2}{L^2} \tag{2.5}$$

Here, q is the drain discharge rate per unit surface area (m/d) in relation to a recharge rate such that recharge is equal to the discharge rate; H is the water table height above drains at midpoint between drains (m); D is the height of the water level in the drains above the impervious floor i.e. thickness of aquifer below drain level (m); L is the drain spacing (m); K_a is the hydraulic conductivity of the layer above drain level (m/d) and K_b is the hydraulic conductivity of the layer below drain level (m/d). It may be noted that equation is derived assuming that the drains are placed at the interface of the two layers and $K_b > K_a$.

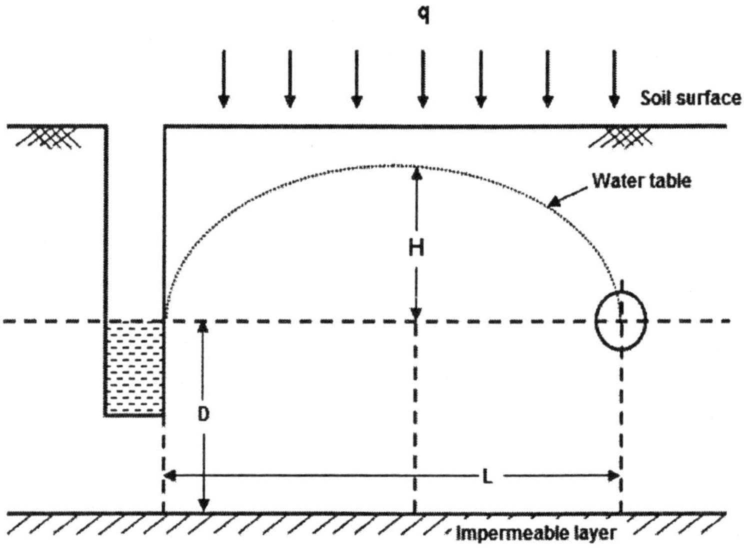

Fig. 2.8 Steady state water table position in drained lands

If the drains do not reach the impervious floor, the flow lines will not be parallel and horizontal but will converge towards the drains (radial flow) causing lengthening of the flow lines and consequently more than proportional loss of head. As a result, the elevation of the water table will be higher where the vertically walled drains are replaced by pipe drains, the drain level remaining the same. To account for the extra resistance caused by radial flow, Hooghoudt introduced a concept of equivalent depth to reduce the depth D to a smaller equivalent depth d_e. The equivalent depth d_e is a function of L, d and r_o (radius of drain). The values of d_e for various values of L, D and r_o can be calculated with the help of following equations.

$$d_e = \frac{D}{1 + D/L\{8/\pi \ln(D/r_e) - \alpha\}} \qquad 0 < D/L \le 0.3 \qquad (2.6)$$

and

$$d_e = \frac{\pi D}{8\{\ln(D/r_e) - 1.15\}} \qquad D/L > 0.3 \qquad (2.7)$$

where,

$$\alpha = 3.55 - 1.6D/L + 2(D/L)^2 \qquad (2.8)$$

and

$$r_e = r_o + \text{fthick} \qquad (2.9)$$

such that r_e is the effective drain pipe radius and fthick is the filter thickness.

By replacing D with d_e, eq. (2.5) can be rewritten as:

$$q = \frac{8K_b\, d_e H + 4\, K_a\, H^2}{L^2} \qquad (2.10)$$

This equation is commonly used to calculate the lateral drain spacing L, if other factors are known. As d_e is a function of L, the use of the formula involves a trial and error procedure. To avoid the trial and error, a number of nomographs have been prepared so that drain spacing could be calculated with minimum complexity. In spite of the availability, application of nomographs has been very limited. With availability of table top computers/calculators, it is easy to calculate drain spacing using simple programs than to depend on the nomographs.

Non Steady state solutions

Most commonly used non steady state solution to work out lateral drain spacing in cases where drains do not rest on the impermeable layer is that of Glover as reported by Dumm (1954). The schematic view of the water table for a non-steady state is shown in Fig. 2.9. The solution for homogeneous soils and for no recharge is given as:

$$L = \pi \left[\frac{KDt}{f}\right]^{1/2} \left[\ln 1.27 \frac{h_0}{h_t}\right]^{1/2} \qquad (2.11)$$

Here K is the hydraulic conductivity of the soil, f is the drainable pore space, t is the time to bring water table height above the drain level from an initial level of h_o to the desired value h_t. The equation is valid for $d >> h_o$.

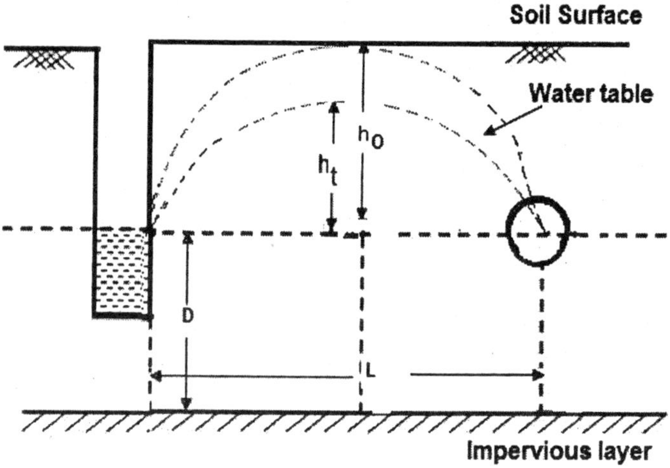

Fig. 2.9 Water table drawdown in drained lands as anticipated in the derivation of non-steady state equations

Dumm (1964) argued that in drained lands initial water table position would not be flat as assumed in deriving eq. (2.11). Assuming the initial shape of the water table as a fourth degree parabola, a drainage design equation has been derived, which is very similar to eq. (2.11). It differs only in the value of the constant 1.27 (4/π) which is now replaced with 1.16. The equation is known as the USBR (United States Bureau of Reclamation) equation and is very commonly used for drainage design in the USA. The exact equation is written as:

$$L = \pi \left[\frac{KDt}{f} \right]^{1/2} \left[\ln 1.16 \frac{h_0}{h_t} \right]^{1/2} \tag{2.12}$$

Glover-Dumm eq. (2.11) and eq. (2.12) do not take into account the radial resistance of the flow towards drains not reaching an impervious layer. To account for the convergence of the flow in the vicinity of the drains, the thickness of the aquifer, D is replaced by d_e as in the case of the Hooghoudt equation. With this change, eq. (2.12) is written as follows.

$$L = \pi \left[\frac{Kd_e t}{f} \right]^{1/2} \left[\ln 1.16 \frac{h_0}{h_t} \right]^{1/2} \tag{2.13}$$

Another modification to eq. (2.13) has been suggested by Tapp and Moody (Dumm, 1964). It takes care of flow from the region above the drains. According to this modification, d_e in eq. (2.13) should be taken as $(d_e + h_o/2)$ or the average initial depth of water bearing stratum $(d_e + (h_0+h_t)/4)$.

In using Glover equation or the USBR equation for drainage design, a design criterion is first decided. The usual design criterion for the Indian condition is to lower the water table by 30 cm in 2 days time once the water table reaches the soil surface. Recently, it has been suggested to increase the time from 2 to 2.5 days. It would result in larger drain spacing and help to save investments.

In addition to the water table and salinity control, some other factors may also play a vital role in deciding drain spacing. For example, the Haryana farmers prefer to have drains on their farm boundaries. After the land consolidation, the farms have attained a regular rectangular shape of 60 x 67 m. Therefore, the drain spacing is either 60 or 67 m depending on the orientation of the drains.

Based on experimental results in India, guidelines on drain spacing as per the soil conditions are given in Table 2.20.

Table 2.20 Guidelines on lateral drain spacing

Soil texture	**Spacing of drains (m)**
Light textured	100-150
Medium textured	50-100
Heavy textured including Vertisols	30*-50

* For spacing less than 50 m, it is desirable to conduct rigorous financial analysis (See Chapter 7)

Size of Lateral/Collector Drain Pipe

Manning's or Darcy Weisback equations are used to determine the size of the laterals/collectors. Considering the capacity reduction due to silt deposition and pipe deflection etc., areas that can be drained with different slopes and different diameter pipes can be computed with a capacity limitation of 60% for pipes less than 100 mm in diameter and 75% for pipes for more than 100 mm diameter (Ritzema, 1994). For corrugated PVC pipes

$$Q = q*A = 2.0 * 10^6 * d^{2.67} * i^{0.5} \quad \text{if } d <= 100 \text{ mm} \quad (2.14)$$

$$Q = q*A = 2.5 * 10^6 * d^{2.67} * i^{0.5} \quad \text{if } d > 100 \text{ mm} \quad (2.15)$$

The corresponding equations for rigid smooth PVC pipes derived on the basis of Darcy-Weisback equations are:

$$Q = 4.6*10^6 . d^{2.714} i^{0.572} \quad \text{for} \quad d \leq 100 \text{ mm} \quad (2.16)$$

$$Q = 5.79*10^6 . d^{2714} i^{0.572} \quad \text{for} \quad d \geq 100 \text{ mm} \quad (2.17)$$

Here, Q is the drainage volume (m^3/d), A is the drainable area (m^2) equal to length of the pipe multiplied by drain spacing, q is the drainage coefficient (m/d), d is the internal diameter of the pipe (m) and i is the slope of drain line (m/m). It may be noted that the maximum length of the lateral increases with internal diameter of the pipe and slope of the lateral. The minimum recommended grade for various diameter pipes are given in Table 2.21.

Table 2.21 Minimum grade, i for tube drain

Drain size (mm)	Grade (%)
100	0.10
125	0.07
150	0.05

The computed pipe diameters are to be rounded off to the next higher commercially available size. In case of clay and cement concrete tiles the minimum diameter prescribed is 10 cm, while for PVC and PE pipes, it is 7.5 cm.

Installation of Subsurface Drainage System

Based on the field investigations and crop drainage requirement, a drainage system is designed and a layout plan is prepared to suit the field boundaries. The topography of the project site receives due consideration in deciding the layout plan. For the installation of drainage system the collector and lateral drain lines are staked and the location for outlet and junction boxes are identified in the field. The installation procedures of some main components of the horizontal subsurface drainage system are as follows:

Outlet

The execution of drainage system begins with the construction of an outlet which could either be gravity or a pumped outlet. The gravity outlets are preferred over the pumped outlets because these are cheaper to construct and do not involve operational cost. If the natural slope permits, the disposal of drainage water from the collector or main drain into an open water body should be considered.

In flat or gentle sloping areas where the gradients are generally insufficient for free fall; it may be difficult to dispose the water by gravity. Under these conditions, a sump is constructed in between the subsurface collector drain and the open channel. The sump provides space for storage of water flowing from the collector drain, before it is pumped into the open watercourse.

Drainage Sump

In monsoon type of climate in India, the ideal period to construct a sump is May or June when the ground water is deepest. A circular pit of a diameter about 1 m more than the diameter of the proposed sump is dug up to the ground water level. At the bottom of the circular pit an iron ring is placed to serve as cutting edge over which a RCC *kerb* is constructed to support the masonry wall to a height 1 m above the ground. The wall is plastered from inside and outside and is provided with holes of 1 cm diameter at regular intervals. The sump is sunk into the ground by removing the soil from the bottom in case shallow water table conditions prevail. Simultaneously, the wall of the sump is also raised so that the sump is 2 to 2.5 m deeper than the depth at which collector drain has been laid. Plastic-coated iron bars are used as steps to go down for cleaning/inspection. In waterlogged sandy soils, quick sand conditions may interfere with the installation of sump. Such situations can be tackled by simultaneously dewatering the area all around the sump while sinking the structure.

Fig. 2.10 A photo of monolithic pipe for sump (left) and collector pipes being readied for installation by the drain trencher (right)

Once the construction of the sump is over, a gravel layer of about 0.5 m thick is placed at the bottom of the sump to serve as a filter for the ground water that flows into the sump. Holes are provided in the side walls for connecting the collector drains. Pre-fabricated reinforced cement concrete (RCC) pipes with monolithic base plate and an inner diameter of 1.2 m for sump and 0.9 m for manhole going to a height of 5 m made of elements of 2.5/1.25 m could also be used to construct sump and manholes (Fig. 2.10 left). A sump of 3.5 m to 4 m inner diameter and 4 m deep below the ground could serve an area of about 40 ha.

Collector and Lateral Drains

According to the drainage plan, the location of the collector and lateral drains is first marked on the ground. The construction work is initiated during a part of the year when the water table is deep enough to facilitate the earth work. The digging of trenches begins from the outlets, so that the system can be kept free of water during rains etc. The trenches for the collector drains should be prepared first. After preparing the trenches for the collector drain, the bed slope is thoroughly checked for any reverse slope. Rigid PVC pipes, stone-ware pipes, non-perforated or perforated corrugated pipes can be used as collector drains. Perforated corrugated pipes, if used, need to be provided with a suitable envelope material in sandy and loamy soils. It may be mentioned that perforated PVC pipes are used in Haryana Operational Project with a 60 mesh nylon sock as the envelope material (Mohan Lal *et al.*, 2012, Fig. 2.10 right). Water in the trenches should be pumped while laying the pipes lines. More so when rigid PVC pipes are used as PVC pipes being light in weight, would float on the water. In such events, it would be difficult to maintain the desired slope.

For lateral drains, once the desired bed slope has been prepared, the gravel envelope is spread in a 30 cm strip at the bottom. The pipe is then laid and is covered with 10 cm thick gravel layer. Process of putting the envelope at the bottom of the trench and above the drains is quite important as it would determine the performance of the drainage system. The drainage pipes and the envelope need to be set in a stabilized soil surrounding. The pipes should be laid in such a manner that formation of soil slurry is avoided. Shallow water table during installation could also adversely affect the performance of the envelope as sharp hydraulic gradients are developed resulting in high inflow velocities which could carry soil particles into the drain. When installation is made under such conditions, even a short-term exposure to high water table conditions can effectively clog the openings of the filter wrapped around the pipes. The chances of sediments flowing into the drains are high during installation and/or during first leaching phase. Once the drains become operational, such high hydraulic gradients might not develop.

Currently, laser controlled pipe laying machines are used to construct drainage systems in Haryana and Punjab and tractor drawn pipe laying machine in Maharashtra (Fig. 2.10 right). The mechanization of drain laying obviates many of the difficulties that are encountered in manual construction of drainage systems.

MONITORING AND EVALUATION

After a horizontal drainage system has been laid out, it is customary to monitor and evaluate its performance over a period of time. Monitoring of the system is essential to achieve the following:

- To assess the adequacy of assumptions made and parameters used in design
- To assess the hydraulic performance of the system
- To assess the benefits of the system
- To prepare guidelines for extending the design to adjoining areas.

The data and the frequency at which data are collected are reported in Table 2.22. Besides what is stated in this table, once in 2-3 years initially and 5 years after that, a portion of the representative drains may be excavated to assess the condition of the drainage materials, the drain pipes and the envelope. Tests on these materials should be conducted to ascertain the strength and hydraulics of the envelope material etc.

Table 2.22 The data and frequency of their measurement

Data	Frequency of measurement
Discharge of individual laterals	Daily in the first season, weekly later on
Discharge from the whole area	Continuously with stage level recorder
Water table midway between drains	Daily in the first season and then weekly
Water level on the drains and 0.5 m away from the drain	Two drains each with different filter materials for one season
Soil salinity	1st year twice, then once in a year, up to 120 cm depth; preferable layers are: 0-15, 15-30, 30-60, 60-90, 90-120 cm.
Quality of drainage effluent	Weekly in the first year and then monthly
Crop yields	Every season for at least 5 years
Drainage installation and land development cost	Keep regular records during installation
Cost of crop production and returns	Every season

Since monitoring and evaluation is a long-term activity and is to be used to prepare guidelines, the data collected should be properly classified, codified and stored in a computer for future retrieval and analysis.

MAINTENANCE: SUBSURFACE DRAINS

Pipe drains being costly to install, these must work efficiently for a long time to get proper returns on investments. Therefore, proper post-installation maintenance is of utmost importance. Only a well maintained system will effectively control the water table and perform other functions for which it is designed and installed. A buried drain very soon goes 'out of sight and out of mind'. Hence a regular inspection and cleaning schedule must be followed. Although, compared to an open ditch system, pipe drainage system needs little maintenance, newly constructed system should be closely observed during the initial one to two years of operation. After the trench has stabilized and most subsidence has taken place, only periodic inspection would be required. A pipe drain requiring attention can be identified by the presence of seepage or soft areas over the drain, sink holes, manholes full of fine material, water backing up in the manholes, shrubs or weeds growing over the drain, iron sludge in the manholes or at the outlet etc. Once these are observed, measures should be implemented without loss of time.

SPECIAL DRAINAGE TECHNOLOGIES

Mole Drainage

Mole drains are subsurface drains without tiles/pipes. These are used when natural drainage is restricted due to lack of slope or heavy clay sub-soil prevents downward drainage. Since water moves so slowly under these situations, successful drainage requires very closely spaced drainage systems, of the order of 2 to 10 m. It would be too costly to use pipe drainage in such situations. Open ditch systems too would be impractical. Mole drains are formed in clay subsoil with a mole plow with a cylindrical foot, often with an expander which helps compact the channel wall. Typically the foot diameter is 75 mm, the expander diameter is 85-100 mm, the leg thickness is 25 mm and the leg length is 200 mm (Fig. 2.11 left). The mole plow designed at Central Institute of Agricultural Engineering (CIAE), Bhopal had a leg with 1000 X 250 X 25 mm and a foot of 100 mm with 110 mm bullet or expander diameter. With a 3 point linkage, the plow can be attached to a wheeled tractor (Fig. 2.11 Right). The foot and expander form the drainage channel and the leg generates the slot with associated soil fissures, which extend from the surface down into the channel. The leg fissures are vertical and are formed at an angle of about 45° to the direction of travel (Fig. 2.12 left). Any surplus water above moling depth can therefore, move rapidly through these fissures into the mole channel. Because mole drains are only formed in the soil, they deteriorate over time. Life and success depends on working with the correct soil type and installing the mole drainage at the right moisture content and at right depth/spacing. It is always useful to provide a PVC pipe at the mole outlet (Fig. 2.12 right). For practical and economic purposes, life of mole channels is usually taken as five years.

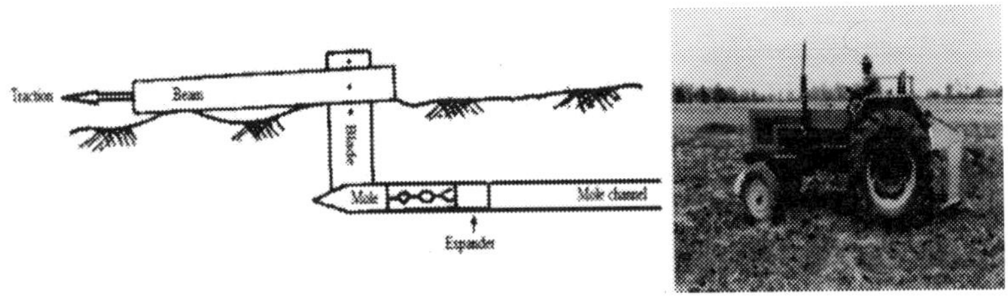

Fig. 2.11 Parts of a mole plow (Left) and a tractor mounted 3 point linkage mole plow (Right)

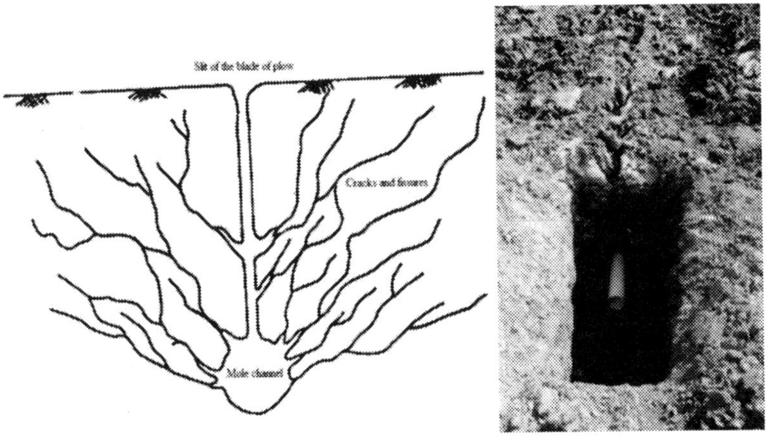

Fig. 2.12 A schematic view of cracks that develop during moling (Left) and a plastic pipe at the outlet end of a mole drain

Soils suitable for moling

As a general guide, the soil at moling depth should have minimum clay content of 30%. The soil must also be uniform across the paddock.

If there are lenses of soil with clay content less than this value at moling depth then moling is unlikely to be successful. Many soils with a clay percentage higher than the minimum do not hold stable moles because they shrink and swell excessively between wet and dry cycles or because they are structurally unstable. At the time of moling, the soil at moling depth should be plastic (moist enough to be moulded by hand into a ribbon without breaking but not wet enough to become liquid when worked in the hand). When moisture content at the mole draining depth is 20-25%, conditions are satisfactory. Besides, for channel longevity, the mole channel should remain dry for at least a month after installation allowing the soil to ripen before the mole conducts water.

Mole drains are generally installed at depths between 400 and 700 mm below the surface although deeper depths up to 1.2 m have been used. The mole drains should be deep enough so that these are protected from the loads exerted by heavy farm machinery and from the swelling and thawing effect of Vertisols. The spacing of mole drainage is usually in the range of 1.5 – 10.0 m. Length of mole drains vary from 20 to 100 m but can be up to 500 m depending on the grade, which may range from nearly level to 5%.

Mole drainage experiences

Mole drains were tried at Digod/Nanta in the Chambal command area in the late seventies under a UNDP project. The mole drains were laid at depths ranging from 0.6-0.8 m and at spacing of 2-5 m covering an area of 17 ha. Recently, Rao *et al.* (2009a) tried this technique for Vertisols of MP at Bhopal with clay and silt content of 55% and 31% respectively. Mole drains 60 m in length were installed at an average depth of 0.6 m at a longitudinal gradient of 0.20%. Three different spacing (2, 4 and 6 m) were used. The cost of mole drains was assessed at 1080, 760 and 640 per ha for a spacing of 2, 4 and 6 m respectively. Mole drains installed in November, 2003 were functional for the next 5 monsoon seasons. The drain discharge varied from about 2 mm to 12 mm depending upon the rainfall. The average soybean (JS-335) yield obtained in 2, 4 and 6 m spacing was more than double in all the treatments as compared to control where no drainage was practiced for all the 4 years of experimentation. Since no significant difference in the yields under 2 and 4 m drain spacing was observed, it emerged that for the Bhopal region a mole drain spacing of 4 m could be recommended (Rao *et al.*, 2009 b).

Vertical Drainage

The removal of ground water by means of a borehole (or well) either under pressure (artesian conditions) or through pumping is termed as vertical drainage. It controls the ground water table or artesian ground water conditions through a group of tube wells having sufficient overlap of their individual cones of depressions. This ensures uniform control of the water table at all points in the area. The vertical drainage system practically includes one or more deep/shallow tube well(s) and the system of open or closed drains to points of disposal. Except for one study on vertical drainage carried out during the first decade of the 20th century, no real long-term vertical drainage study has been conducted in India. One/two studies conducted in Punjab and Gujarat have been functional only for a short period due to operational and maintenance problems. Yet, the system performance could be judged from the success achieved in tackling the problems of water logging that was prevalent in sixties in parts of Punjab, Haryana and Uttar Pradesh.

Calculation of number of tube wells for different options

In the design of vertical drainage system, it is imperative to calculate the number of tube wells required to control the water table. Three scenarios *viz.* control of water table to avoid damage to crops; vertical drainage-cum-irrigation and vertical drainage on the basis of replenishable ground water recharge are considered. For the three situations, formulae to calculate numbers of tube wells are given in Table 2.23.

Guidelines to decide number of tube wells

Decision on the actual number of tube wells to be installed on the basis of 3 numbers obtained is crucial for long-term sustainability of hydraulic regime of the area. Judicious decisions should be taken deriving support from local site specific situations. The following guidelines may also be considered:

- The number of tube wells should be installed on the basis of irrigation requirement of the crops or on the basis of pumping out the net replenishable recharge, whichever is high. If numbers of tube wells are more for irrigation, the net pumping should be restricted to net replenishable recharge unless some lowering of the water table is intended.
- If it is intended to lower the water table for short periods, the designed tube well system should be run for longer number of

hours to lower the water table at a faster rate than the designed duration.

Table 2.23 Formulae to calculate number of tube well for water table control

Scenario	Formula	Variables
Control of water table to avoid damage to crops	$N = \frac{w_p \times 24}{C \times n}$	N= Number of tube wells needed to drain the area w_p= (A x 0.3 x f)/2 A = Area m^2 f= Drainable pore space C= Tube well discharge m^3/d n= Number of hours tube wells run in a day 0.3 and 2 are used to lower the water table by 30 cm in 2 days
Number of tube wells for vertical drainage-cum-irrigation	$N = \frac{C \times n \times m}{I \times 24}$ N= A/Ac	I= Depth of irrigation, m m= Irrigation interval Ac = Area commanded by one tube well All other parameters are same as in row 1
Vertical drainage on the basis of seasonal replenish-able ground water recharge	$N = \frac{A \times Np \times 24}{C \times n \times n_1}$	Np= Replenishable recharge, m n_1= Average number of days a tube wells would run in a year All other parameters are same as in row 1

Vertical drainage experiences

In the year 1987-88, a pilot vertical drainage system was installed in about 700 ha area by the Department of Irrigation and Drainage, Govt. of Punjab at Gulabewala in Muktsar block of Faridkot district (Sondhi, 2007). Fourteen tube wells (30 m deep) were installed in 5 rows with a row-to-row spacing of 1000 m and well-to-well spacing varying between 71 to 94 m (Fig. 2.13). In addition, 10 piezometers were installed to monitor the effect of pumping on water table. The pumping began in July, 1987. From July, 1987 to February, 1988, depending upon the availability of power, the average working hours of the tube wells varied form 2.67 h/d to 23.47 h/d. The fall in water table during this period ranged between 0.08 to 0.51 m.

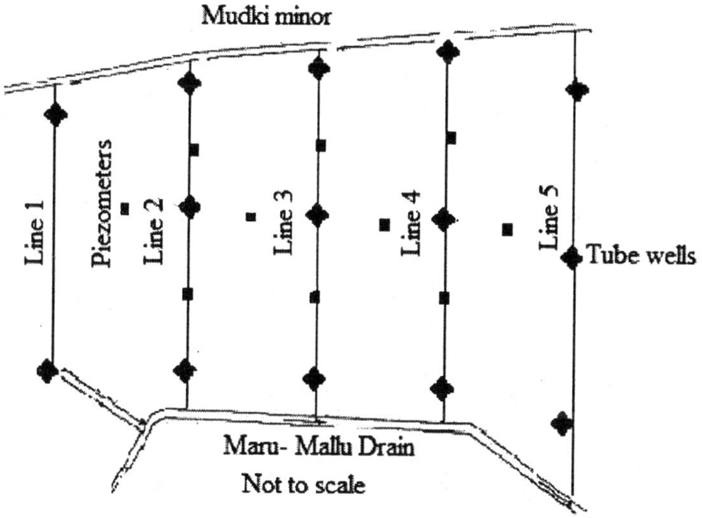

Fig. 2.13 A Layout of vertical drainage system at Gulabewala in Punjab

Since these tube wells were not designed considering the vertical variation in ground water quality, a marked change in water quality in the project area was observed during July, 1987 to January, 1988. The EC values increased from 2.04 to 3.35, 6.40 to 7.34, 4.18 to 4.71 and 5.14 to 5.56 dS/m for line 1, 2, 3 and 4 respectively. Not much change was observed in line 5. This increase in EC values was due to upconing of saline water body situated at a deeper depth. Apparently any vertical drainage scheme must look into the tube well water quality.

Bio-drainage

Biological drainage or bio-drainage essentially means the application of a biological material to drain water stagnating on the land surface or to lower the water table. Quick growing plants transpiring at a high rate are usefully employed as bio-drain materials. Bio- drainage plants should have the following characteristics.

- The plants should have the ability of deep rooting to lower the water table to greater depths
- The plants should be able to use significant amounts of water over all the seasons of the year

- The plants should be able to withstand water logging and salinity/alkalinity
- The plants should be able to adapt to widely varying climatic conditions and moisture supply. It should also be resistant to diseases
- The plants should have a ready market so that it should be possible to easily dispose the produce.

Fast growing cloned *Eucalyptus,* known for luxurious water consumption under excess soil moisture condition, is suitable for bio-drainage. Other suitable species for block plantations are *Casuarina glauca, Acacia ampleceps, Terminalia arjuna (Arjun), Pongamia pinnata (Papri)* and *Syzygium cuminii (Jamun)* etc. (Dagar, 2011). Recent studies indicate that maximum lowering of water table is mainly restricted to planted areas (bio-drained) only. The zone of lowering of the water table extends only up to 10 m away from the planted areas.

To work out the area to be planted with given specie of the tree, following steps are followed.

- Calculate the water balance of the area to assess the extra water to be disposed
- With known depth to water table, the salinity of the ground water and for the soil texture of the region, assess plant transpiration to open pan evaporation ratio (T/E_{pan} ratio) for the kind of plants proposed to be planted
- Calculate annual water use by the trees using T/E_{pan}
- Calculate percent area to be planted by dividing excess water by annual water use of the plant.

If the percent area to be planted is more than the desired limit, either change the bio-drainage plant or integrate conventional techniques with bio-drainage.

Limitations of bio-drainage

Two major limitations of the bio-drainage are: loss of precious land and water and loss of crops yield due to shadowing effect or damage caused by birds/insects and pests. Another major limitation is due to plants inability to absorb harmful salts. In this situation, while water table can be controlled to a limited extent, it would be difficult to control soil salinity (Gupta, 2005). In several studies it has been observed that salts in a bio-drained field accumulate at a relatively deeper depth than in the controlled area where salts accumulate at the soil surface. It is a positive feature of bio-drainage.

Bio-drainage experiences

From a lysimetric study, bio-drainage rate up to 5.5 m/annum for *Eucalyptus tereticornis* and up to 4.2 m/annum for *Bambusa arundinacea* have been reported. (Chhabra and Thakur,1998a, Chhabra and Thakur, 1998b). Plant growth was adversely impacted by depth and salinity of the ground water and so was the bio-drained water, which also reduced with depth and salinity of the ground water. No salt uptake by the trees was observed yet the salt remained lower in the profile as compared to unplanted lysimeters.

The evapotranspiration rates were much less under field situation than the lysimeteric studies. In three years old plantations, the evapotranspiration rates were 29.5, 19.8 and 14.4 L/d/plant in low, optimum and high densities, respectively which comes to 189, 339, 945 mm during May, 2003 to December 2003. The average evapotranspir-ation values increased to 56.5, 30.7 and 18.9 L/d/plant in respective categories from August, 2005 to December 2005, (Minhas, 2005). After six years, annual total consumptive use of water was 2200 mm under high density and 1300 mm in low density plantation, which seems to be quite reasonable. Wood production is affected adversely when high loading rates are used to dispose the water (Dagar, 2011).

The parallel strip plantations of *E. tereticornis* raised on *killa* lines to lower the shallow water table at Puthi and Baas villages in Hisar district revealed that the ground water table underneath the plantation strips was lower than the adjacent fields. During April, 2005, when trees were two years and three months old, the average draw down under plantation was 18 cm which increased to 85 cm during April 2005 to April 2008. The spatial extent of lowering of water table in the adjacent fields was observed even beyond a distance of 66 m from the edge of outer strips (Dagar, 2011). Farmers could harvest a total wood biomass (timber suitable for poles/*ballies* and pulp wood) of 36 t/ha from 5 years and 4 months old trees. The total above ground biomass of timber, fuel wood, twigs and leaves was 39 t/ha and the below ground biomass of extractable roots was 15 t/ha. The plantation could sequester 18 tonnes of carbon/ha in above ground and 7 tonnes of carbon/ha in root. Benefit-cost ratio at a discount rate of 12% of first rotation (5 years and 4 months) was 3:1 against 1.3:1 for agricultural crops in Haryana. It would be many-fold for the next 3 to 4 rotations as the maintenance cost of coppiced *Eucalyptus* would be negligible. Currently, this system of drainage is being extended to larger areas in the Jhajjar District, Haryana (Fig. 2.14).

Fig. 2.14 A view of bio-drainage at Paharipur in Jhajjar district

Failure of the conventional open interceptor drains at many places in India and Pakistan helped to strengthen the belief that bio-drainage could be an attractive option to intercept the seepage by planting trees alongside the water carriers (Manjunatha *et al.*, 2000; Lal, 2005). Field results have revealed that *Acacia nilotica, Dalbergia sissoo, Sesbania grandiflora* and *Casurina equisetifolia* intercepted 86, 84, 72 and 72% of the canal seepage (Patil *et al.,* 2005).

REUSE/DISPOSAL OF DRAINAGE WATER

Saline/alkali drainage water is a major byproduct of horizontal and vertical drainage systems. One of the major issues that confront a drainage engineer is safe disposal of the drainage water. The disposal and/or its reuse must ensure that there are no adverse effects on the soil and water resources of the region and the environment at the disposal site remains undisturbed. Both the quality and the quantity of drainage water would influence the selection of safe means of reuse/disposal. One of the plus features of subsurface horizontal drainage over the vertical drainage system is that in the former, quantity of drainage water is much less and therefore, the disposal problems would be minimal. Secondly, water quality in the deeper strata is usually poor. Thus, pipe drainage offers more favorable opportunities to reuse drainage water as compared to vertical drainage.

Drainage Water Quality

The temporal variation in the quality of drainage water from a pipe drainage system, the quality of water from a cavity well (20 m depth) and the canal water at the experimental site at Sampla generally attest to the statement made in the previous paragraph (Table 2.24). The data also reveal that there is significant improvement in the quality of drainage water over time. This observation has now been firmed up from several other experimental sites.

Table 2.24 Average composition of drainage water, ground water and the canal water during irrigation at Sampla

Water	EC (dS/m)	SAR	Solute concentration (meq/L)			
			Ca+Mg	Na	HCO_3	Cl
Drainage water						
1986-87	26.5	16.8	173	156	1.8	315
1987-88	27.0	17.0	187	164	2.2	325
1988-89	18.0	14.7	135	121	2.1	232
1989-90	15.0	14.5	126	115	1.6	221
Ground water (20m)	32.5	11.9	280	141	4.7	405
Canal water	0.7	0.8	2	0.8	2.0	0.8

Characterization of Water

Depending upon the EC, SAR and RSC of water, drainage water could be categorized into three main classes (Table 2.25). Saline and alkali water have further been sub-grouped into three classes each taking into account the severity of the problems. Similar to soil classification, alkali and sodic terms are synonymously used to describe alkali waters. Toxic water, a special category of water, has excess of one or the other specific ions (Table 2.25)

Table 2.25 Grouping of ground water for irrigation in India

Water quality	EC_{iw} (dS/m)	SAR_{iw} $(mmol/L)^{1/2}$	RSC (meq/L)
A. Good	< 2	< 10	< 2.5
B. Saline			
i. Marginally saline	2-4	< 10	< 2.5
ii. Saline	> 4	< 10	< 2.5
iii. High-SAR saline	> 4	> 10	< 2.5
C. Alkali water			
i. Marginally alkali	< 4	< 10	2.5 – 4.0
ii. Alkali	< 4	< 10	> 4.0
iii. High-SAR alkali	variable	> 10	> 4.0
D. Toxic Water	The toxic water has variable salinity, SAR and RSC but has excess of specific ions such as chloride, sodium, nitrate, boron, fluoride, or heavy metals such as selenium, cadmium, lead and arsenic etc.		

Management of Drainage Effluents

Per Se saline/alkali drainage water do not affect the crop production but the continuous use of such water makes the soil saline and/or alkali that causes the plants to suffer when irrigated with saline/alkali waters. As such, most technologies that are appropriate to manage saline and alkali lands would also be valid when the lands are irrigated by saline/alkali drainage waters. But, even then many strategies that pertain to the use of saline/alkali water have been developed and grouped under crop management, land management, irrigation water management, chemical management and rainwater management (Fig. 2.15). With the use of one or a combination of more than one strategy, drainage waters could be profitably used in agriculture. Elaborate information on the technologies depicted in Fig. 2.15 has been developed and could be accessed from the literature (Minhas and Gupta, 1992; Manchanda *et al.*, 1989, Bajwa *et al.* 1998, Gupta *et al.*, 2002, Gupta and Gupta, 2002, CSSRI, 2004, Gupta, 2011). The relative tolerance of crops to saline water is given in Annexure I, which could be used to select crops for use of saline drainage effluents. Based on extensive and intensive studies, guidelines for the use of saline/alkali drainage waters have been developed and are reported in Table 2.26 and Table 2.27.

The alkali hazard of the drainage water on the soil can be mitigated by neutralizing the RSC of irrigation water with gypsum provided the

RSC of the irrigation water is more than 2.5. Since the application of chemical amendment is to increase the calcium content of the soil, amendments containing calcium viz., gypsum, calcium chloride and calcium polysulfide have been used for alkali water management. The quantity of pure gypsum needed to neutralize one meq/L of RSC comes to 86 kg/ha per 10 cm of water. Accordingly, agriculture grade gypsum (70% purity) @ 12 kg/ha should be added for each cm of alkali water applied. The quantity of gypsum to be added is determined by the RSC of irrigation water and quantity of water required for irrigation during the growing season or on yearly basis.

$$G_r = \frac{86 \times (Ca)_g \times N_i \times D_i}{P_g \times 10000} \qquad (2.18)$$

G_r is the amount of gypsum required (t/ha), (Ca)g is the difference of the RSC of irrigation water to the desirable RSC, D_i is the depth of each irrigation (cm); N_i is the number of irrigations and P_g is the purity of gypsum (%).

Management of Saline/alkali Water

Crop Management	Crop selection Exploitation of varietal differences Growth stage sensitivity Agronomic practices
Soil Management	Soil type specific management Land formation/Seeding Fallowing
Irrigation water Management	Shallow depth-high frequency irrigation Conjunctive use Pre-sowing irrigation Post-sowing irrigation Swithover to improved irrigation techniques Leaching
Chemical Management	Application of additional nutrients Amendments application
Rainwater Management	

Fig. 2.15 Management of saline/alkali drainage water

Table 2.26 Guidelines for using saline waters for different crops

Soil texture (% clay)	Crop tolerance	Upper limits of EC_{iw} (dS/m) in rainfall regions (mm)		
		< 350	350-550	550-750
Fine	S	1.0	1.0	1.5
(<30)	ST	1.5	2.0	3.0
	T	2.0	3.0	4.5
Moderately fine	S	1.5	2.0	2.5
(20 – 30)	ST	2.0	3.0	4.5
	T	4.0	6.0	8.0
Moderately coarse	S	2.0	2.5	3.0
(10 – 20)	ST	4.0	6.0	8.0
	T	6.0	8.0	10.0
Coarse	S	-	3.0	3.0
(< 10)	ST	6.0	7.5	9.0
	T	8.0	10.0	12.5

S, ST and T denote sensitive, semi-tolerant and tolerant crops to salinity of irrigation water

Special Considerations for Table 2.26

- Use gypsum when saline water has SAR > 20 and/or Mg/Ca ratio > 3 or is rich in silica or causes surface stagnation of water during rainy season and crops grown are sensitive to water stagnation.
- In conjunctive use mode, canal water should be used at early growth stages including pre-sowing irrigation.
- If saline water is used for pre-sowing irrigation, 20% extra seed rate and a quick post-sowing irrigation ensures better germination.
- In low rainfall areas, it would be appropriate to leave the fields fallow during the rainy season when SAR > 20 and waters of high salinity are used to irrigate the crops during the *rabi* season.
- Additional application of phosphorus fertilizers should be applied when Cl/SO_4 ratio is > 2.0.
- Use of organic materials in saline environment enhances yields.

Comments

- Textural criteria should be applicable for all soil layers down to at least 1.5 m depth.
- In areas where ground water table reaches within 1.5 m at any time of the year or a hard sub-soil layer is present in the root zone, the limit of the next finer textural class should be used.
- If in any textural class, SAR and RSC values are above the upper limits, gypsum should be used as per the requirement.

Table 2.27 Guidelines for using alkali waters

Soil texture (% clay)	Upper limits of		Remarks
	SAR	RSC	
Fine (> 30)	10	2.5 – 3.5	• Limits pertain to *kharif* fallow/*rabi* crop rotation when annual rainfall is 350-550 mm.
Moderately fine (20-30)	10	3.5 – 5.0	• When the water has Na < 75% (Ca + Mg > 25%) or rainfall is >550 mm, the upper limit of the RSC range becomes safe.
Moderately coarse (10-20)	15	5.0 – 7.5	
Coarse (< 10)	20	7.5 –10.0	• For double cropping, neutralization of RSC with gypsum is essential

Source: Minhas and Gupta (1992)

Disposal Strategies

The possible options for disposal of drainage water are: disposal of drainage water into river system; collection and storage in evaporation tanks; disposal into deep aquifers and disposal to the Sea. Besides cultivation of tree and grasses, economically exploitable halophytes, aquaculture and prawn culture could also be used to dispose of the water and at the same time get reasonable profits. Attempts have also been made to prepare salts from highly saline drainage water although its economics has so far not been established.

3

SOIL SALINIZATION AND LEACHING

Secondary soil salinization of agriculturally productive irrigated lands has assumed a serious dimension. Large areas in major and medium irrigation projects have either gone out of cultivation or productivity has reduced so much that the farmers are unable to make a living out of it. Timely recognition of salt accumulation in the root zone therefore, has become a major issue to prevent secondary salinization. Although, high irrigation efficiencies are sought to obviate the adverse effects of shrinking water resources and to minimize drainage water disposal related problems, yet with reduced leaching fraction, risk of salinity build-up could be much more than in the present situation. As the demand for fresh water in other sectors of economy grows, agriculture gradually has to switchover to poor quality water. As such, one has to regularly monitor the root zone so that soil productivity remains at the optimum level.

In any scheme of reclamation of salt affected soils, high ESP or the high salt concentration in the root zone must be brought to an acceptable level before cropping could begin. For this, one time leaching, a process through which salts accumulated in the root zone could be reduced to an acceptable level is an essential pre-requisite. Besides this, leaching requirement must be met to maintain salt balance of the reclaimed root zone especially when salty water is used for irrigation. Considering these issue, salt balance components, processes operative during soil salinization and desalinization and salt leaching methods have been discussed in this chapter.

SALT BALANCE OF THE ROOT ZONE

The water and salt balance of the root zone are so intricately woven that it is not possible to separate them out. The salt balance of the root zone of irrigated lands can be worked out by various components of water balance (Fig. 3.1) multiplied with their respective salt concentrations (Gupta and Khosla, 1996). The resulting salt storage or leaching from the root zone can be mathematically put into the form of an equation as follows:

$$S = R\,C_r + I\,C_i + G\,C_g + S_1 - DC_d - S_2 \quad (3.1)$$

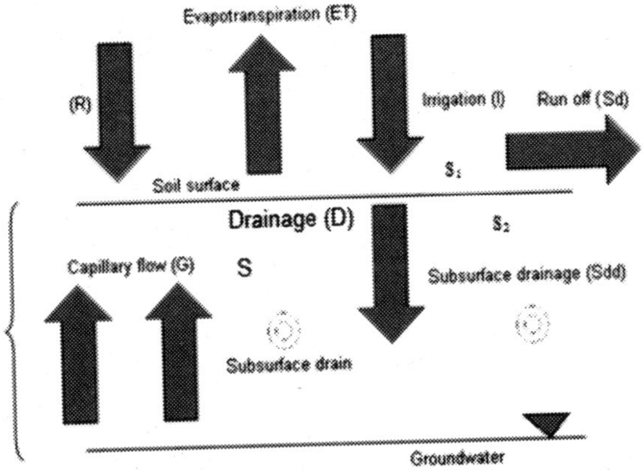

Fig. 3.1 Water balance components of the root zone of irrigated lands

Here salt storage (+) or leaching (-), S, in the root zone is given by the components which add salt to the root zone (+) or those which remove the salt (-). These components are added together with their respective signs. In eq. (3.1) R, I, G and D are amounts of rainfall, irrigation, ground water contribution and drainage below the root zone and C is the concentration of water with subscripts describing the respective components. In this equation salt are added to the root zone through rainfall (RC_r), application of irrigation water (IC_i), ground water (GC_g) and direct addition of salts such as through the application of amendments/fertilizer or generated in the root zone through mineral dissolution (S_1). On the minus side are the salt removed due to drainage (DC_d) and the salt precipitation (S_2). In most cases, the terms S_1 and S_2 are neglected leaving only four terms on the right hand side of eq. (3.1). It may be noted that irrigation induced mineral dissolution is significant and it may not be possible to neglect this component in irrigated lands without causing significant differences in the calculated and actual salt balance. It is clear that to control salinity build-up in the root zone one has to induce drainage from rainfall and applied irrigation water. It could also be achieved by minimizing the salt addition to the root zone. Since rain water is a good quality water, the salt additions could only be controlled either by reducing the salts applied with irrigation water or through the control of ground water which might add substantial amounts of salts in the root zone if ground water is at shallow depths and/or is saline in nature. Such tasks are easier said than done. For example, application of irrigation water or the contribution of the ground water to meet the crop water requirement cannot be controlled at will as it may adversely affect the productivity due to inadequate supply of water causing matric stress. Moreover, unknown parameters in eq. (3.1) add another dimension to the complexity of eq. (3.1). For example G and C_d are not exactly known. Therefore, it is always useful to monitor the salt status of irrigated lands over a few years time to work out the salt regime. It would not only help to diagnose the problem but would also identify factors likely to accentuate salt accumulation or leaching of the salts.

Factors Affecting Salt Balance in the Root Zone

Soil salinity being dynamic, has large spatial and temporal variability. Some of the factors that influence the salinity of the root zone are the amount of water applied, the salinity of the irrigation water, depth to water table, soil texture and other soil characteristic which affect its permeability, evapotranspiration, total rainfall and its distribution, irrigation methods employed, setting of the land in respect of the adjoining area (topography) and above all the overall management inputs that goes with any agricultural activity. The following salt balance scenarios could be anticipated:

- Balance in favor of leaching (S in eq. (3.1) is negative)
- Stable salt balance (S in eq. (3.1) is zero)
- Balance in favor of salt accumulation (S in eq. (3.1) is positive)

Temporally, salt balance is determined for short durations (say between irrigations or a month), medium durations (say a season) and long durations (say annual) over a few years time. Similarly, salt balance is established over small terrestrial units (say a farm) or over an irrigation command or on a regional basis. Usually, larger areas are chosen in salt balance studies, since salt moves from one area to another. Salt transport from an irrigated area to an unintended area is schematically shown in Fig. 3.2. Such processes operating on regional scale generally justify investments on assessing regional salt balance along with smaller units. While preparing the salt balance sheet, causes, rate and degree of accumulation or leaching needs to be established.

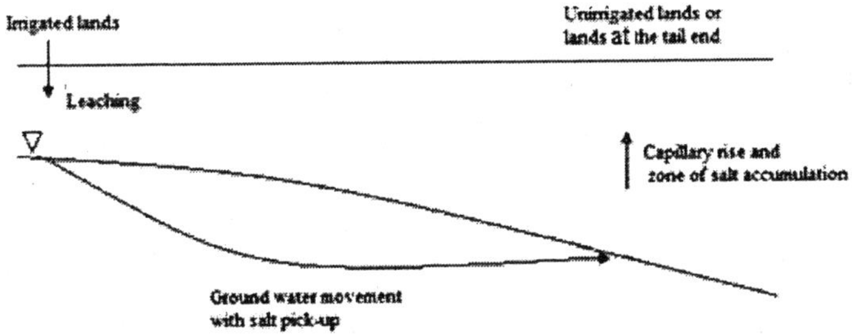

Fig. 3.2 Salt movement in irrigated lands showing areas of salt leaching and accumulation

Soil Salinization of Irrigated Lands

Soils, which are otherwise salt free, may be rendered unproductive by irrigation water containing excess amount of salts. In irrigated lands particularly when limited quantities of water is applied, the salts accumulate in the root zone if equivalent amounts are not removed through appropriate means. The salinization-desalnization cycle in a saline water irrigated land in monsoon climatic conditions shown in Fig. 3.3 reveal:

- The period of soil salinity build- up (Zone A) during November to April till equilibrium salinity is reached
- Period of salt redistribution (Zone B) within the root zone from April to mid June or till the onset of monsoon. During this period, soil salinity is usually maximum in the surface layer (0-15 cm).
- Period of desalinization (Zone C) during monsoon season when rainfall is in excess over the potential evapotranspiration.

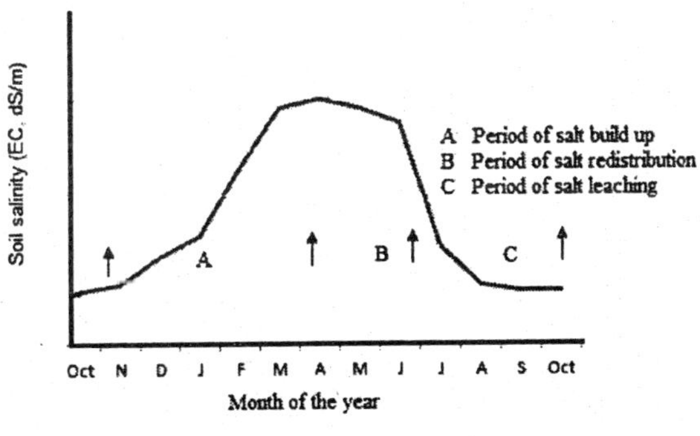

Fig. 3.3 Salinization-desalinization cycle in a saline water irrigated land

Several factors control the extent of salinity development in the soil profile, five being the most important. These are (a) quality of irrigation water, (b) number of irrigations, (c) soil texture (d) leaching fraction and (e) rainfall.

Quality of irrigation water

Soil salinity increases with increasing salinity of irrigation water. If 10 cm of water having EC = 0.5 dS/m is applied to irrigate an area, it would add about 320 kg/ha of salts. If the EC is 1 dS/m, the addition would be doubled to 640 kg/ha. The relationship between the soil salinity (EC_e) and the quality of irrigation water (EC_i) is expressed in the form of a regression equation:

$$EC_e = a + b\ EC_i \qquad (3.2)$$

Where, 'a' and 'b' are regression coefficients. The regression coefficients vary from one location to another (Table 3.1) and might show large variations in both the coefficients. It could be due to variations in soil texture, number of irrigations, leaching fraction and rainfall or even variations in the management practices which together have great influence on the soil salinization-desalinization cycle besides the irrigation water quality.

Table 3.1 Regression constants of eq. (3.2) for four stations

Station	Regression constants	
	a	b
Agra[1]	1.59	1.02
Dharwar[1]	0.50*	0.24
Jobner[2]	3.87	0.48
Pali[2]	0.84	0.44

[1]Data from experimental stations. [2] Data from farmers fields (*) Regression constants for EC_2 versus EC_i

Number of irrigations

The regression constants of eq. (3.2) for the Agra station show that the constant 'b' varies from a low value of 0.66 for two irrigations to 1.21 for five irrigations (Table 3.2). The average value for 3-5 irrigations is 1.02 while for 2-3 irrigations is 0.83. It may be noted that if all factors remain unaltered, there would be little change in soil salinity after 5-6 irrigations. A regression equation which relates soil salinity with irrigation water quality and number of irrigations is of the following form:

$$EC_e = -2.26 + 0.90\ EC_i + 1.24\ N. \qquad (3.3)$$

Here, N is number of irrigations (Gupta, 1986). Equation (3.3) is applicable to medium texture soils where number of irrigations does not exceed six.

Table 3.2 Regression coefficient as influenced by number of irrigations

Number of irrigations	Crop	Regression constants of eq. (3.2)	
		a	b
2	Wheat	1.54	0.66
3	Wheat	2.14	0.99
4	Wheat	2.07	1.00
5	Wheat	0.62	1.21
3-5	Wheat	1.59	1.02
2-3	Mustard	2.00	0.83

Soil texture

Build-up of soil salinity is more in heavy than in light textured soils. Under similar situations, the regression coefficient, b, varies from 0.36 for sandy to loamy sand group of soils to 0.69 for fine loam, clay loam and silty clay loam group of soil. The value of 'a', may not vary much and is in the range of 2 to 2.2 for the two groups of soils.

Leaching fraction (LF)

Leaching fraction is defined as the fraction of the irrigation water that leaves the root zone. It can easily be known from the average water application efficiency. An application efficiency of 70% means that leaching fraction is around 0.30. In general, the steady state salinity in the soil could be expressed by the following equation:

$$EC_e = EC_i * X \qquad (3.4)$$

Here X is the concentration factor, which depends upon the leaching fraction. It may be noted that eq. (3.4) is similar to eq. (3.1) except for the parameter a, which is zero in eq. (3.4). The relevant concentration factors as controlled by leaching fraction are reported in Table 3.3. The data show that at low leaching fraction the steady state salinity could be about 3.2 times the salinity of the irrigation water while at high leaching fraction it could be even less than the salinity of the irrigation water as is evident from leaching fraction exceeding 30%. The steady state root zone soil salinity (EC_e) likely to develop upon long-term use of saline water at various leaching fractions (LF) and a water uptake pattern of 40:30:20:10 is shown in Fig. 3.4. The water uptake pattern of 40:30:20:10 means that from the four quarters of the root zone from top to bottom, applied water is used by the crop in the ratio of 40:30:20:10 (Fig. 3.5). The uptake pattern could vary from one crop to another and due to soil layering. The example given herein illustrates the calculation of the concentration factor, X for a leaching fraction of 0.2. The steady state salt profile likely to develop in this case with two water uptake patterns is given in Fig. 3.5.

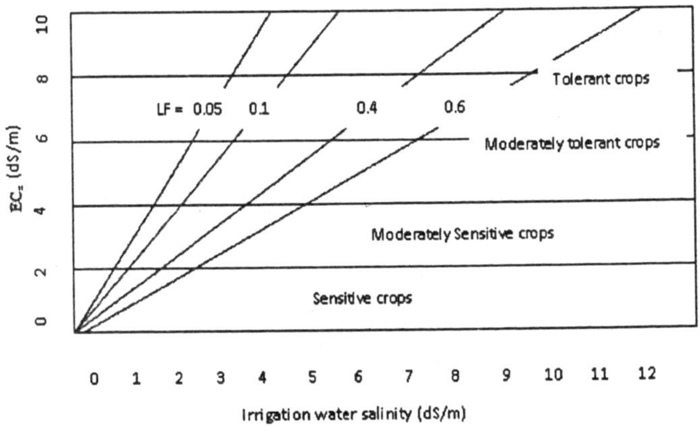

Fig. 3.4 Steady state root zone soil salinity (EC_e) likely to develop with applied irrigation water salinity (EC_i) at various leaching fractions (LF)

Table 3.3 Concentration factor (X) for use in eq. 3.4

Leaching fraction (LF)	Applied water needed (% of ET)	Concentration factor (X)
0.05	105.3	3.2
0.10	111.1	2.1
0.15	117.6	1.6
0.20	125.0	1.3
0.25	133.3	1.2
0.30	142.9	1.0
0.40	166.7	0.9
0.50	200.0	0.8
0.60	250.0	0.7

Source : Ayers and Westcot (1985)

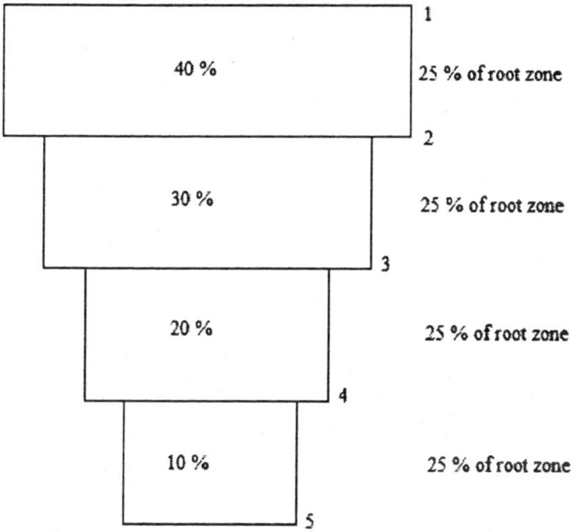

Fig. 3.5 Water uptake pattern by the plants

Example: A crop is irrigated with a water of ECi 4.0 dS/m maintaining a LF of 0.2. Calculate the average root zone salinity assuming the water uptake pattern of 40:30:20:10.

EC_1 (Fig. 3.5) = 4.0 / 1.0 = 4.0/1.0 = 4.0 dS/m

EC_2 =4.0/(1-(0.40) (1-0.20)) = 4.0/0.68 = 5.88 dS/m

EC_3 = 4.0/ (1-(0.40+0.30) (1-0.20)) =4.0/0.44 = 9.09 dS/m

Similarly

EC_4 = 4/.28 = 14.28 dS/m

EC_5 =4/0.2 = 20.00 dS/m

Thus EC_{sw} can be calculated as

EC_{sw} = (4.00+5.88+9.09+14.28+20.00)/5 =10.65 dS/m

or $EC_e = EC_{sw}/2$ = 10.65/2= 5.32 dS/m

Concentration factor = EC_e/EC_i = 5.32/4.0 =1.33 times of EC_i

In Table (3.3), value at this LR = 1.3

EC of drainage water, EC_{dw} EC_i/LF = = 20.0 dS/m

Concentration factors for other water uptake patterns could be calculated following the same procedure. For example, for a 60:25:10:5 uptake pattern, the concentration factor is 1.52 instead of 1.33.

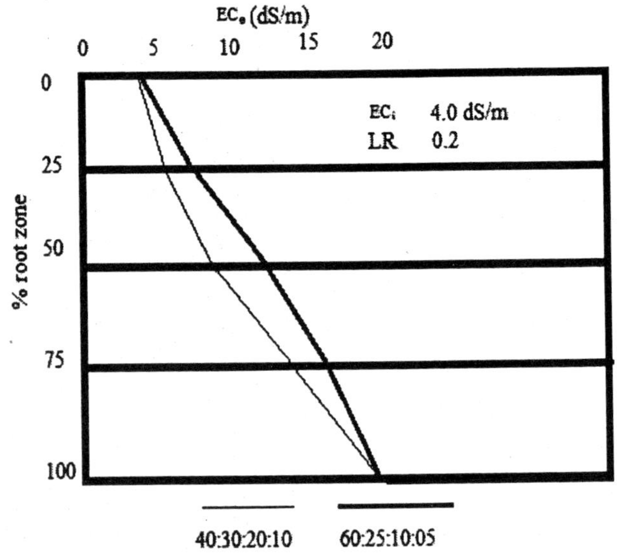

Fig. 3.6 Steady state salt profile in a saline water irrigated land with two uptake patterns

Equation (3.4) could also explain the effect of soil texture on soil salinity build-up. For the same amount of water applied, the leaching fractions will be more in a light than a heavy textured soil resulting in higher salinity build-up in the later. Apparently, the effect of amount of irrigation water applied and the soil texture could be lumped together in leaching fraction. Since the mean root zone salinity could be as low as half and as high as 4 times the salinity of the irrigation water, there is ample scope to control the mean root zone salinity at a pre-decided level by controlling leaching fraction. It may be noted that in monsoon climatic conditions, steady state conditions do not prevail and hence the concept of leaching fraction has little relevance under the Indian monsoon climatic conditions.

Rainfall

Rainfall has twofold effect on soil salinity build-up. On the one hand, salt load is reduced through reduced number of irrigations and on the other rainwater being very good quality water, helps to leach the salts accumulated in the root zone. With an additional 10 cm of rainfall, the constant `b' got reduced by about 0.6 units for wheat and mustard at Agra (Table 3.4).

Table 3.4 Effect of rainfall on regression constants of eq. (3.2) at Agra

Crop	Rainfall (mm)	Number of irrigations	Regression constants	
			a	b
Wheat	17.1	4	2.44	1.29
Wheat	113.9	3	1.76	0.69
Mustard	18.2	3	2.76	1.11
Mustard	113.9	3	1.80	0.50

Soil Alkalization

Sodium adsorption ratio (SAR) determines the sodium hazard of irrigation water especially when no precipitation of calcium carbonate is expected. It should be noted that an increase in soil water concentration causes SAR to increase in proportion to the square root of the concentration factor.

$$SAR_{\text{soil water}} = SAR_i * \sqrt{n} \tag{3.5}$$

Here n is soil water concentration factor. Since part of water in the soil solution is provided by precipitation, SAR of the soil water is usually less than that calculated using SAR of the irrigation water alone. It has emerged that SAR_{i+p} is around $SAR_i * \sqrt{0.7}$, although exact factor may depend upon the rainfall and the irrigation water applied.

A number of issues have been raised on the applicability of SAR to assess alkali hazard of the water but no one opinion has emerged so far. For example, inclusion of Mg in the calculation of SAR has been questioned by many experts. According to this group, adverse effects of Mg are not as serious as that of Na and are not as good as that of Ca. Since its effects are intermediate of the two ions, clubbing of Mg with Ca has not found favor. Under Indian conditions where Mg/Ca ratios tend to increase with salinity of water and are more than one in many cases, clubbing of Ca and Mg may underestimate SAR hazard of water. The only justification to club calcium with magnesium appears to be that Ca and Mg were reported together in the past. Since calcium plus magnesium is divided by 2, one may not worry much about the Mg till the Mg/Ca ratio is around 1.0. Gupta (1984) suggested that when Mg/Ca ratio is more than 1.0, Mg in eq. (1.6) could be omitted and only calcium could be used to calculate SAR and designated as SCAR. He used this concept to predict ESP of the soils, which better correlated with ESP developed in the soils than SAR calculated with the help of eq. (1.6) (Table 3.5). In spite of many such suggestions, no clear alternative solutions have emerged so far.

Table 3.5 ESP/SAR of irrigated soils as related to SAR and SCAR

Location	EC_i (dS/m)	Mg/Ca ratio	SAR_i	SCAR	Irrigated soil	
					SAR	ESP
Dantiwada	3.8	3.2	16	24	17	18
Kaparda	10.8	4.4	28	59	31	60
Jelwa	5.9	8.1	23	36	28	38
Surayata	9.0	5.4	8	16	6	14

If precipitation is expected, then Besides SAR, residual sodium carbonate (RSC) is calculated to assess whether calcium carbonate is likely to precipitate or not. A high RSC is a warning signal that the SAR of the soil solution may strongly increase exceeding the normal increase proportional to the square root of the concentration factor. Eaton (1950) also used soluble sodium percentage (SSP) and SSP (possible) to describe the hazard of water that is high in bicarbonate.

$$SSP = (Na * 100)/ (Ca + Mg + Na) \tag{3.6}$$

$$SSP\ (possible) = (Na * 100)/ [(Ca + Mg + Na)-(CO_3 + HCO_3)] \tag{3.7}$$

In eq. (3.7) deduction part should not exceed (Ca + Mg). Here also questions have arisen whether magnesium should be clubbed with calcium as magnesium does not precipitate as magnesium carbonate (Suarez, 1975).

Few equations have been proposed to take care of precipitation and dissolution of Ca. Ayers and Westcot (1976) proposed adj. SAR, which is calculated by the following relation:

$$adj.\ SAR = SAR\ (1 + 8.4 - pH_c) \tag{3.8}$$

Such that pH_c in eq. (3.8) is calculated from eq. (3.9) with (Ca + Mg) as was proposed by Bower (1961).

$$pH_c = (pk_2 - pk_c) + p(Ca+Mg) + p(Alk) \tag{3.9}$$

The tests proved that eq. (3.8) over predicts the sodium hazard and should be multiplied by a factor of 0.5 to evaluate more correctly the effect of HCO_3 on calcium precipitation (Ayers and Westcot, 1985). As such, the use of this equation is not recommended to calculate the sodicity hazard of irrigation water.

Suarez (1981) proposed another equation to assess the dissolution and precipitation of calcium, and designated it as adjusted R_{Na}. The equation is written as follows:

$$adj.\ R_{Na} = \frac{Na}{[(Ca_x + Mg)/2]^{1/2}} \tag{3.10}$$

Here Ca_x is the value of Ca in the water that takes care of EC_i, HCO_3/Ca ratio of the water and the estimated partial pressure of CO_2 in the surface few mm of the soil (Annexure II, Table A-1). The Ca_x values represents meq/L of Ca expected to remain in the soil solution at equilibrium. The calculation procedure is also illustrated in Annexure II.

Some workers introduced the concept of leaching fraction in some of the equations realizing that precipitation of $CaCO_3$ is likely to decrease with increasing LF. Thus, adj. SAR equation is modified to (Bower, 1961).

$$\text{Adj.SAR} = \left\{\frac{1}{\text{LF}}\right\}^{1/2} \text{SAR}_i \,\{1 + (8.4 - \text{pH}_c)\} \tag{3.11}$$

Where LF is the leaching fraction. Rhoades (1968) further modified the equation as follows:

$$\text{SAR}_{dw} = \frac{Y^{1+2LF}}{(LF)^{1/2}} \text{SAR}_{iw} \,\{1 + (8.4 - \text{pH}_c)\} \tag{3.12}$$

Here Y is an empirically determined mineral weathering coefficient normally taken as 0.65. Rhoades (1968) observed that eq. (3.11) would predict the ESP of the soil surface and eq. (3.12) the ESP of the soil at the bottom of the root zone. Clearly, ESP at the soil surface is more than that at the bottom. Since eq. (3.8) over predicts the sodium hazard by a factor of 0.5, it is likely that eq. (3.12) would also over predict the adverse effects being similar to eq. (3.11) at LF =1.

Minhas and Sharma (2006) indicated that Adj. R_{Na} concept is relatively better provided it is modified to take care of the dilution and dissolution processes under field conditions. As such, they modified eq. (3.10) by considering a concentration factor, CF and a simple dissolution factor, DF as follows:

$$CF = (Dlw/ET) + (Dlw/Drw)$$

$$DF = Drw/Dlw$$

Here, Dlw is the amount of alkali water applied (mm), Drw is the annual rainfall and ET is the annual crop evapotranspiration. Eq. (3.11) is modified to:

$$\text{Adj. } R_{Na}\,(\text{mod.}) = (CF/DF)^{1/2} * \text{Adj. } R_{Na} \tag{3.13}$$

or

$$\text{Adj. } R_{Na}\,(\text{mod.}) = (Dlw/Drw)\,(1+Drw/ET)^{1/2} * \text{Adj. } R_{Na} \tag{3.14}$$

In spite of the limitation as revealed through various developments in this section and tests failing to validate the universal application of even the modified equations, use of SAR given by eq. (1.6) or eq. (1.7) and RSC given by eq. (1.8) are still practiced widely to characterize and arrive at management options for use of poor quality water. There is a need to go deeper into the issues raised to arrive at a more theoretically sound characterization scheme for poor quality alkali ground water.

Prediction of Soil ESP

One must be able to predict the soil ESP with continuous use of alkali water to initiate corrective measures against the build-up. Most equations relate soil ESP with SAR of the water. On the contrary, Indian experiences have revealed that if SAR of the water is high due to neutral salts, build-up of ESP might not occur to an extent that it may prove detrimental to the crops. Moreover, high SAR of the soil solution caused by neutral salts gets reduced with leaching. Nonetheless, tests have revealed the usefulness of these equations although rigorous tests are needed to come to a firm conclusion under Indian conditions where monsoon climatic conditions prevail.

According to USSL (1954) Handbook, the relationship between SAR of the water and soil ESP is give by the following equation:

$$ESP = \frac{100\,(-0.0126 + 0.01475 * SAR_i)}{1 + (-0.0126 + 0.01475 * SAR_i)} \qquad (3.15)$$

Since SAR of the soil water could be 1 to 3 times more than that of the irrigation water, higher ESP build-up may be expected than given by eq. (3.15).

Bower and Maasland (1963) used eq. (3.11) to predict the sodium hazards of ground water of Punjab (Pakistan) as follows:

$$ESP = 2\ SAR\ \{\ 1+(8.4 - pH_c)\} \qquad (3.16)$$

Factor 2 is obtained by taking LF=0.25 such that $1/(LF)^{1/2}$ is equal to 2.

Bajwa *et al.* (1983) reported that observed soil ESP was quite close to SAR_{dw} (Rhoades, 1968) but the differences between the observed ESP and predicted SAR increased with increase in SAR_{adj} of the irrigation water. Bajwa *et al.* (1992, 1998) compared eq. (3.11) and eq. (3.12) equations for their capability to predict irrigation hazard with sodic and saline-sodic water in a cotton-wheat rotation. It was observed that modified Ayers and Westcot (1985) equation predicted the developed

ESP for all sodic and saline-sodic (high RSC) water but none of the models predicted the build-up of soil ESP for high SAR and low RSC water for which SAR_{dw} equation performed better.

In a recent paper, Minhas and Sharma (2006) used data of large number of experiments including the above referred studies to develop and test eq. (3.14). Following observations have been made:

1. Adj. R_{Na} concept (Ayers and Westcot, 1985) suitably modified to take care of concentration factor that is dependent upon applied alkali water, evapotranspiration and annual rainfall and dissolution process is able to predict the hazards of alkali waters. Predicted ESP using the modified adj R_{Na} concept could be given by ESP = 0.96 * adj R_{Na} (mod). Thus, it is possible to establish one to one relationship between ESP and the sodicity parameter adj R_{Na} (mod), which has been eluding so far.
2. It is also observed that the threshold limit of ESP tolerance of paddy/millet is relatively less when ESP builds-up with the application of alkali water compared to when the ESP reduces from an initially high value with gypsum application as in an alkali land reclamation program (Table 3.5). It seems that adverse effects would be more pronounced in alkali water irrigated lands than for lands under reclamation (Minhas and Sharma, 2006).

Table 3.5 Comparative tolerance of crops to soil ESP in alkali water irrigated crops and in alkali land reclamation

Crop	Threshold ESP	
	Alkali water irrigated land*	**Soil reclamation#**
Wheat	16.2 (1.86)	16.4 (2.1)
Paddy	20.1 (1.62)	24.4 (0.9)$
Millet/maize	6.1 (1.34)	13.6 (2.6)
Cotton	14.9 (1.26)	NA

*Minhas and Sharma (2006); # Gupta and Sharma (1990); Values in parenthesis are the slope of Mass and Hoffman (1977) equation; $ Values may not be used to extend beyond ESP_{50} i.e. ESP at which yield is 50% of the optimum; NA = Not available

Minhas and Sharma (2006) using eq. (3.14) predicted the adj. R_{Na} of the irrigation water that could sustain the yield of various cropping sequences under different rainfall conditions (Table 3.6).

Table 3.6 Predicted values of adj. R_{Na} of irrigation water for sustaining yield under different cropping sequences

Cropping sequence	Permissible adj. R_{Na} for sustainable yields in rainfall zones (cm)		
	<40	40-60	>60
Fallow-wheat	16	21	27
Maize/millet-wheat	14	17	23
Paddy-wheat	6	9	14
Cotton-wheat	14	20	26

Source: Minhas and Sharma (2006)

Soil Salinization due to Capillary Movement

Mineral soils are derived largely from the weathering of rocks. A close relation exists between salts in the soil and the chemical composition of rocks from which these soils are formed. Soils formed from salt bearing rocks usually contain more salts. In some cases, the salts under natural condition lie at some depth, but aided by man's activities get transported into the root zone.

Ground water and capillary rise

Ground water in arid and semiarid regions often lies at deeper depths usually in equilibrium with natural recharge and drainability of the land. At such depths, there is no fear of water transport due to capillary rise. Once, this equilibrium is disturbed with the introduction of irrigation, the water table starts rising until it reaches within the capillary fringe. At that time, water evaporates at the land surface, leaving behind salts in the root zone turning the land saline/alkali. The total quantity of water brought into the root zone and the salt concentration of the ground water determine the extent of soil salinization during the fallow period. The amount of salts brought in to the root zone is calculated by the following equation:

$$S = C*m \tag{3.17}$$

Here S is the amount of salt added, kg/ha, C salt concentration of ground water, g/L or kg/m^3 and m is amount of capillary flow entering the targeted zone in m^3/ha. For the change in EC_e of the soil due to added salts, following equation could be used.

$$\Delta EC_e = \Delta S/0.7*w_e*\rho_b/\rho_w*D \tag{3.18}$$

Here w_e is the water content in mass percentage, D is the depth of the root zone, cm and ρ_b.and ρ_w are the bulk density of soil and density of water respectively i.e. 1000 kg/m^3. The factor 0.7 is used to convert C to approximate EC dS/m.

If 400 m^3/ha water having 14 g/L salts enters the root zone the total salts that enters the root zone is calculated with the help of eq. (3.17) which comes to 5600 kg/ha. Assuming root zone depth as 50 cm, w_e as 50% and ρ_b/ρ_w as 1.5, the change in EC of the root zone can be calculated by using eq. (3.18) and it comes to 2.1 dS/m for this case.

Critical Depth to Water Table

The capillary rise/flow is related to soil texture, depth to water table and water gradients. While capillary rise increases with increasing silt content of the soils, the rate of evaporation increases with decreasing depth to water table (Fig. 3.7). The concept of 'Critical Depth to Water Table', first advocated by Talsma (1963), has been in vogue for quite some time. According to this concept, the process of secondary salinization is quite active once the water table is above this critical depth. The critical depth, for the purpose of this concept, is defined as the depth to water table at which relative evaporation is 0.1 or the rate is about 10% of its value when the water table is at or near the soil surface. For practical standpoint, the evaporation rate should not exceed 0.5 mm/d to 1 mm/d, the maximum allowable flux to maintain favorable water and salt balance. On this basis, critical depth may vary from a meter in light textured soils to a few meters in fine textured soils.

Recent studies have revealed the limitations of this concept and arguments contesting the applicability of this concept have been put forth. It is argued that if the flux at the bottom of the root zone is kept downward, there would be no opportunity for soil salinization even under shallow water table situation. In the Indian context, subsurface drains installed at shallow depths performed well and maintained water and salt balance because flux of the water at critical periods is downward (monsoon season, when the water table is at shallow depths). During the summer season, when evaporative demand is high, water table is deep enough to allow excessive upward flux. Thus, applicability of the concept of critical depth to water table needs to be examined before it is applied to field situations. Otherwise, we may end up in costly solutions in land reclamation programs.

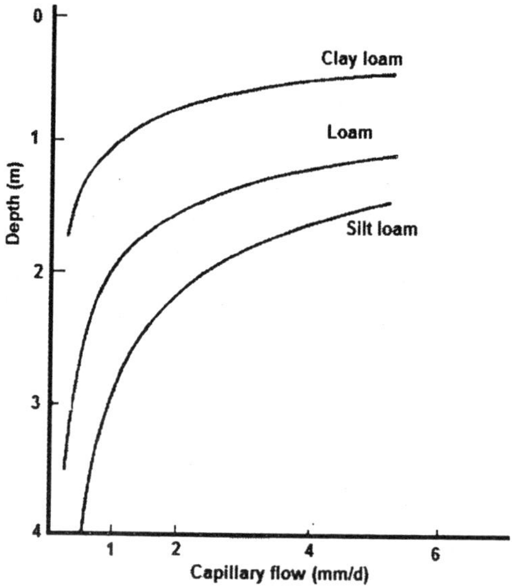

Fig. 3.7 Capillary flow velocity and depth of water table in 3 texture soil groups (van Hooren, 1979)

Permissible Maximum Flux

Water is the main carrier of salts in the soil. As the soil solution reaches the site of evaporation at or near the soil surface, the water is evaporated and the salts are left behind. In cropped lands, the zone of salt build-up also depends upon the depth and the distribution of active roots, which may transpire a part of the water even before it reaches the soil surface. Salinity in the profile increases with decreasing depth to ground water table but increases with increasing salinity of the ground water at the same depth to water table. For favorable water and salt balance, understanding and knowledge of the maximum flux, q_{max} is quite important. For good quality ground water, conventionally 1 mm/d is taken as the limiting value but for poor quality ground water, the value should be reduced to minimize soil salinization. The effect of ground water salinity (0.6-3.7 dS/m) was observed to be more pronounced at shallow than at deeper depths. The studies have also shown that correlation between the concentration of the salts at the soil surface and the ground water is significant only when water table is at or above 1.2 m.

The generalized relationship between water table depth and upward water flux suggests that water can move at some appreciable rate from water tables as deep as 5 m. It is also observed that the evaporation rate decreases more steeply with water table depth in a coarse than a fine textured soil. The maximal limiting upward flux (cm/d) is given by the following relationship:

$$q_{max} = \frac{A\,a}{d^n} \tag{3.19}$$

Here d is the depth to water table (cm), a (cm^{n+1}/d) and n are coefficients and A is also a constant which depends on n. Some typical values of these constants are presented along with maximum upward flux in Table 3.7.

Table 3.7 Experimental values of constants of eq. (3.19)

Soil texture	**n**	**a (cm^{n+1}/d)**	**A**	**Maximum upward flux# (cm/d)**
Sand	4	$1.7*10^8$	1.52	2.58
Fine sandy loam	3	$3.2*10^5$	1.76	0.56
Loam	2	$1.7*10^3$	2.46	0.42
Clay	2	$1.1*10^3$	2.46	0.27

Source: Gardner and Fireman (1958); # For d=100 cm

Capillary Rise and Soil Salinity Build-up

Under shallow water table situations, yield is reduced due to poor aeration, limited root extension and salt build-up especially when the ground water is saline. The distribution of salts in the root zone would be entirely different in a saline land formed due to saline water irrigation and a land salinized due to capillary rise. In the former, salts increase with depth while in the later salts decrease with depth being in equilibrium with the subsoil salinity. Nonetheless, techniques which aid the crop to draw water from the shallow water table are employed to reduce the desired number of irrigations, lower the production costs and decrease the drainage volumes and attendant problems. Pump operational schedules in drained lands are designed to manipulate depth to water table to aid the crop to draw water from the shallow water table to meet a part of its crop water requirement. Under such situations, provision of leaching after the cropping season should be made in advance. In the monsoon climate, such opportunities are naturally available due to excess rainfall during the monsoon season.

Time Frame

In irrigated agriculture, time frame within which salt balance of the root zone should be assessed is very important not only to save water but also to have good crop growth. The time frame normally depends upon the salinity build-up. For a salt sensitive crop irrigated with saline water, salt balance may be important following an irrigation event. For salt tolerant crops, it should be possible to allow build-up of the salt in the root zone for two-three irrigations or even for a crop season. In the monsoon climatic conditions where annual salinization-desalinization cycles are common, zero or negative salt balance of the root zone should be aimed at on a yearly basis. Within a year, there could be periods of salt build-up depending upon the tolerance of the crop.

Sustainability of Irrigated Agriculture

In recent times, serious doubts have been raised on the sustainability of irrigated agriculture. Past experiences where civilizations have been ruined are quoted to emphasize the ill effects of irrigated agriculture. Without doubt, past experiences where irrigated agriculture not only failed to bring prosperity but ruined the potential beneficiaries could be cited to sensitize the people, yet there is ample scope to ponder over as to what has gone wrong and why? One could learn from these mistakes and rectify the same in current irrigation projects. Today, we have much more field experiences fully supported by theoretical exercises assuring us that there is no serious threat to permanence of irrigated agriculture. However, this threat could be real, if man fails to use the practices and knowledge now available to him to devise location specific ways and means to fight the menace of environmental degradation. To practice long-term sustainable irrigated agriculture, one could take support of field verified models to generate scenarios to arrive at best interventions that minimize salinity build-up. Any failure of irrigated agriculture at this stage would be the failure of the mankind to put into practice present day knowledge on the subject. In any case, it would not mean the failure of irrigated agriculture.

LEACHING

Definition and Objectives

Leaching of saline soils is essentially a process in which water of low concentration is applied to displace the soil solution of relatively high concentration. The application of the excess water with the aim of pushing the salts below the root zone is defined as leaching. The leaching of salts is carried out to meet one of the following objectives:

- To remove or to reduce the salts in the root zone commonly known as one time leaching for reclamation
- To maintain the salt balance in the reclaimed or irrigated lands so that crops do not suffer due to excess salts at any time in future.

Historical Background

The importance of leaching to reclaim salt affected lands had been recognized long ago. Mild alkali soils were reclaimed by flooding them with water for several months followed by rice transplanting (Leather, 1914; Mehta, 1940). Severely affected salty lands could be reclaimed by impounding water and draining away the salts. Leaching alone of alkali soils, however, failed to produce any response on crop yield or soil conditions (Singh and Nijhawan, 1932). In the southern states, leaching alone with sugarcane crop was not successful (Talati, 1941) while leaching with drainage and green manuring with *dhaincha* (*Sesbania*) had great ameliorating effect (Talati, 1947). *Khar lands* (coastal saline soils) can be reclaimed if protected from tidal inundations and leached by rain water or through artificial irrigations. Kanwar (1962) while reporting the results of trials to reclaim saline soils by leaching alone or in combination with gypsum and fertilizers, concluded that leaching alone is not beneficial. Dutch scientists concluded that continuous ponding did not yield quick desalinization in *Bhal* tract of Gujarat because of reduction in infiltration rates upon impounding of water (NEDECO, 1968). Short inundation cycles were better because infiltration rate remained high throughout. Tilling in between the inundations proved beneficial.

Flushing for Salt Removal

Process of flushing the salts is quite different than leaching in many ways although the aim is same i.e. to reduce salts in the root zone. In the former, salts are removed by surface washing of the soil as compared to pushing of soluble salts below the root zone in the latter. In flushing, salts move with runoff or irrigation water as the water moves down the sloping land surface. The water with washed away salts is removed through a surface drain provided at the tail end. Reeve *et al.* (1955) reported that salts removed by flushing in a 100 m border strip constituted only 1% of the total salt content of the 60 cm deep soil profile.

Yadav and Agarwal (1959) observed that the volume of leachate and the quantity of sodium salt removed from a saline-alkali soil increased when the pots were given flushing in between two irrigations. Gupta (1985) observed that except for the first storm or storms following long dry spells, no appreciable amount of salt is present in the runoff occurring due to heavy rain over saline soils. The flushing method is useful only when soil permeability is low or in soils where impermeable layer exists near the soil surface. Elgabaly (1971) recommended that where a salt crust of highly hygroscopic salts prevents use of leveling or other heavy equipments because of low bearing capacity; it may be useful to flush the salts. It will permit better drying of the surface soil. A combination of flushing and leaching has been successfully used for the reclamation in several projects of *Khar land* Development Board, Gujarat.

Flushing proved effective in reclamation of saline-sodic soils of the central India. For effective flushing, water should be kept ponded on the land surface for 2-3 days followed by puddling for effective dissolution of soluble salts followed by drainage of the ponded water. Two flushing with an interval of about 7 days in between the two flushing is recommended. Since the soluble carbonates and bicarbonates of sodium could be effectively flushed, gypsum application after the flushing proved relatively more effective than without flushing. In the later case gypsum reacts with soluble salts to adversely impact the gypsum efficiency. Recently, it has been established that in relatively heavy textured highly saline-alkali soils of Uttar Pradesh, flushing along with leaching is a better option than leaching alone (Nayak *et al.*, 2008).

If flushing is carried out with a continuous stream of water, the grade in the forward direction should be reasonable for water to move fast. Such grades may pose difficulties in the design of the irrigation system as well as for uniform spread of the water during leaching. Since, a well designed surface drainage system is essential for this technique, flushing may not be practicable in land locked fields. This technique as a means of reclamation has limited application due to its high water requirement, inefficient salt removal and several technical constraints.

Leaching for Reclamation

The only reasonable way to accomplish reclamation seems to be by applying adequate quantities of water to the soil surface, such that the excess salts present in the root zone are dissolved and transported below the root zone with percolating water. For efficient leaching following steps should be practiced.

Field preparation

Natural undulating and unbunded fields should be first bunded and then leveled. Inadequate leveling results in spotted desalinization and may cause delay or failure of the reclamation program. To enhance uniformity of leaching, smaller plots with temporary bunds could be constructed such that these plots are later adjusted within the overall layout plan of the farm. Tillage of land before and in between leaching is beneficial because it would enhance infiltration, reduce the time of reclamation and improve the leaching efficiency.

Time of leaching

Availability of water in adequate quantities is an essential pre-requisite to design a successful leaching program. Whenever such quantities of water can be arranged should be most appropriate time to initiate a leaching program. If there are no constraints of water, the best time for efficient leaching is when the water table is deepest and evaporation is minimal. Leaching of sodium sulphate, however, might be retarded during winter season when evaporation is less. If water table is at shallow depth during winter, it is better to choose a period when water table is at least 1 m deep. If leaching is sought to be accomplished through rainfall, it is natural that one has to wait for a period when excess rainfall is expected. For example, in the Indian context, at most places in arid and semiarid regions, excess rainfall is experienced only during the monsoon season. Therefore, July-August period would be the best time for leaching. If, some water is available towards the beginning of this period, it can be used for leaching as it would enhance the efficiency of leaching through rainfall. Even saline water could be used to increase the moisture content of the profile to enhance leaching by rain. The salinity of water in this case should be less than the salinity of soil solution in the top layer (0-15 cm) of the profile.

Depth of reclamation

The depth of reclamation depends largely upon the potential land use. It is 0.6 m for shallow rooted crops or pastures to more than 2 m for trees. For diversified cropping, the depth of reclamation usually varies from 1 to 1.5 m. It may be noted that these are the minimum recommended depths and appropriate measures against resalinization should be adopted.

How much to leach

The salt status of the soil following leaching is normally determined by the tolerance of the crop to be grown soon after reclamation. The salt constituents of concern (salinity, specific ion concentration and/or alkali) should not exceed the permissible limit. If the level achieved after leaching is insufficient to cultivate the proposed crop(s), either a more tolerant crop should be selected or additional water must be applied to leach the salts. Most comprehensive data on salinity tolerance of crops have been published and are often used to frame guidelines for crop selection in reclaimed lands (Maas and Hoffman, 1977; Maas, 1980). The crop tolerance data given in Table 3.8 provides limited information on some crops for some selected stations in India.

Table 3.8 Crop tolerance to soil salinity

Sites	Soil type	Crop	Threshold (dS/m)	Slope (%)
Sampla	Sandy loam	Wheat	4.0	29.0
		Mustard	6.0	75.0
		Barley	7.0	19.0
Karnal	Sandy loam	*Moong* bean	1.8	20.7
		Mustard	3.8	6.9
		Sorghum	2.2	10.6
Agra	Sandy loam	Wheat	8.2	19.8
		Mustard	6.1	20.7
		Berseem	3.5	12.5
		Potato	4.4	16.1
		Tomato	1.3	6.5
Dharwad	Black clay	Wheat	2.3	20.5
		Safflower	2.8	20.7
		Sorghum (W)	2.1	3.9
		Italian millet	6.1	50.0
		Sataria	1.12	2.19
Indore	Black clay	*Berseem*	2.0	11.22
		Safflower	2.80	5.00
		Maize	0.50	7.90
Chakuly	Clay	Rice	0.60	2.70

Crop productivity at the end of the season depends upon the initial crop stand. Therefore, most favorable conditions must be created during the germination and initial establishment stages. In case of one time leaching for reclamation, amount of salts to be leached is usually expressed as a fraction (Lf) of the amount of salts present in the profile at the start of leaching.Mathematically,

$$\text{Lf} = 1 - \frac{\text{Threshold salinity of the crop}}{\text{Initial salinity of the soil}} \tag{3.20}$$

If some yield reduction is permissible, then threshold salinity in eq. (3.20) could be replaced by the salinity at which desired yield reduction (EC_d) would occur. It can be calculated by the data reported in Table 3.8 utilizing the following relationship:

$$EC_d = \frac{\text{Desired yield reduction (\%)}}{\text{Slope (\%)}} + \text{Threshold EC} \tag{3.21}$$

If tolerance of the crop at the germination stage is less than the mean values in Table 3.8, it would be desirable to use the threshold salinity levels at the germination stage.

Leaching Curves

In planning reclamation programs, the leaching curve is a useful tool to decide whether desalinization is technically and economically feasible and which reclamation method should be adopted. Leaching curves could be developed with sufficient accuracy using simple and rapid leaching tests. While conducting these tests, one should adopt the same process of leaching as is likely to be used in the reclamation project. A simple, straight forward way to prepare leaching curves is to run a test similar to the one used to evaluate infiltration rate. Two concentric cylinders are used for this test (Fig. 3.8). The water level in the inner cylinder is kept at a constant level and water infiltrated is measured till the pre-determined amount of water has entered the soil. The experiment is usually a replicated trial. In these experiments, 3 or 4 depths of water are passed and are taken as the treatments. Soil samples in the inner cylinder are taken before and after the completion of the test to evaluate moisture content and salt status. For salt status appraisal, the most convenient approach is to measure the bulk soil electrolyte conductivity of different layers.

The amount of water drained below a given depth, D_w is found by subtracting from the total depth of water infiltrated, the amount of water evaporated and the amount of water absorbed in the soil layer. The EC_{eq} is commonly used to make the leaching curve independent of the salinity of irrigation water and the existing evaporation and drainage conditions. It is obtained by measuring the soil salinity in the 0-5 cm layer in the treatment where largest amount of water has entered the profile. It is expected that the salinity of this layer reaches a practically constant minimum soon after the leaching commences.

Fig. 3.8 A view of the set-up to measure infiltration rate and to prepare leaching curves

The leaching curve is the dimensionless plot of $(EC_e - EC_{eq}) / (EC_o - EC_{eq})$ (EC_o is the initial soil EC_e) versus D_w / D_s (D_s is the depth of the soil). The data are plotted either on an arithmetic paper (Fig. 3.9 left) or on a semi-log paper (Fig. 3.9 right). These curves are used later on to determine one time leaching requirement for land reclamation for the test soil and conditions under which leaching tests have been conducted. The data show that for the same fraction of initial soil concentration, water required to leach a light textured soil is relatively less than a heavy textured soil (Fig. 3.9 left).

The leaching curves so prepared may not be applicable to many site specific situations such as chemically amended alkali soils and drained lands. Leaching in chemically amended alkali soils is not only required to transport the applied chemical constituent of the amendments to the site of exchange, but also to transport the exchanged ion and soluble salts below the root zone. The leaching efficiency in this case is largely governed by the kind and solubility of the chemical amendment. Commonly, gypsum and pyrites are applied as amendments to reclaim alkali soils. Leaching practice for the two amendments are different and would be discussed in Chapter 4. In drained lands, the main limitation of a leaching curve so prepared relate to the type of flow encountered. In constructing leaching curves, flow is vertical and therefore, leaching curves so prepared will be valid for situations where water movement is vertical or practically vertical. Thus, the leaching curves prepared would not be applicable to strips near the drains or 1 m below the soil surface in widely spaced laterals with lateral drain depths of 1.5 m or more where flow is radial (Batta *et al.*, 1980). Practical approach to leach drained lands has been discussed in Chapter 5.

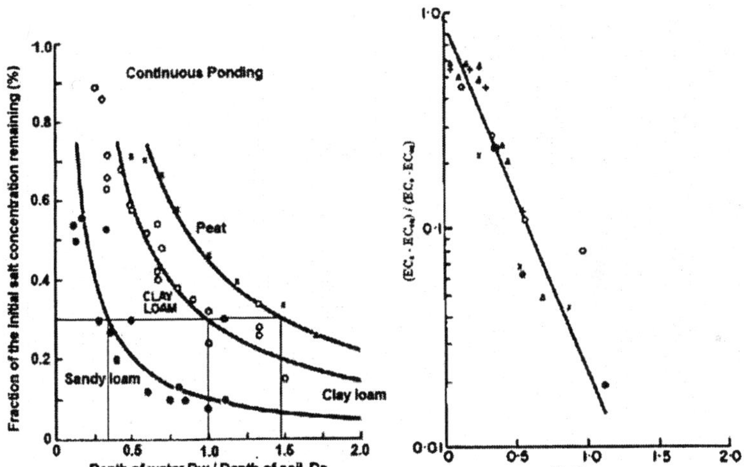

Fig. 3.9 Leaching curves for different soil texture groups (left) and for a medium texture soil (right)

Amount of Water Required for Leaching

In an ideal soil system without bypass, without dissolution or precipitation of salts, without salt diffusion or dispersion, the salt concentration of the soil water passing a given depth in the soil profile should drop to the concentration of the applied water when one pore volume (1 PV) of the applied water equals one pore space or one pore volume of the soil volume to be leached. The actual amount of water required under practical leaching will, however, depend upon many factors. Important among them are:

- Amount of salt initially present in the profile
- Desired level of salt in the root zone
- Soil texture
- Type of salts present
- Soil depth to be reclaimed
- Efficiency of the drainage system, if provided
- Method of leaching

As a thumb rule, a unit depth of water will remove about 80% of the salts from a unit depth of the soil (Reeve *et al.*, 1955). Field data, however, show a wide range depending upon the soil texture and other factors listed in Table 3.9. The amount of water required to reclaim 1 cm of the soil under field conditions varied from a low value of 0.3 cm to a high value of 2.58 cm with continuous leaching (Table 3.9). Very high values reported by Beyce (1972) appear to be a result of greater bypass due to the cracking nature of the peat soil.

Continuous and Intermittent Leaching

Leaching efficiency could be improved by creating unsaturated conditions during leaching and/or by reducing the average rate of water movement. This can be achieved through intermittent ponding of water on the soil surface. Intermittent leaching minimizes loss of water through large pores and provides greater opportunities for diffusion of salts from immobile or less mobile pockets to the mobile zone. It has emerged that as much as 30 to 70% water required for leaching in continuous ponding mode could be saved through intermittent leaching. Nielsen *et al.* (1966) reported that intermittent ponding may not be useful if surface layer is more restrictive than the lower layers. In such cases unsaturated conditions prevail in the profile even under continuous ponding. Intermittent and continuous ponding performed equally well in alkali soils (Abrol and Bhumbla, 1973), sandy loam soil (Lefellar and Sharma, 1977) and for reclamation of saline soils formed as a result of 2.5 years of application of saline water (Jury *et al.*, 1979). Similar results were reported for leaching of boron (Beyce, 1972). Dahiya *et al.* (1981), however, reported that nearly 50% saving in water could be made with intermittent leaching compared to continuous leaching in a sandy loam soil. Elgabaly (1971) reported that although intermittent ponding saved water, it took nearly 40% more time for leaching compared to continuous leaching. Surface evaporation of water occurring between leaching may sometime overshadow the increased leaching efficiency brought about by intermittent leaching. Synthesis of present day data reveal that intermittent leaching may be useful in soils with large bypass (with cracks or structured soils), poor soil permeability, deep ground water table and during periods of low evaporative demand. On the other hand, continuous ponding may give reasonably good results in soils with good permeability, unstructured soils, areas that have shallow saline ground water and during periods of high evaporative demand.

Table 3.9 Amount of water required for leaching as affected by soil type, ion monitored and method of leaching

Soil type	Method of leaching	Constituent monitored	Amount of water applied for leaching (cm/cm)	Salt removed (%)
Deep peat[1]	Continuous	EC	2.58	80
	Intermittent	EC	4.43, 4.17	80
	Intermittent sprinkling	EC	2.25	80
	Continuous	B	3.69	80
	Intermittent	B	6.33	80
	Intermittent sprinkling	B	3.68	80
Silt loam[2]	Intermittent	EC	0.30	*
Loamy sand/loam[3]	Continuous	EC	0.40	*
Sandy loam [4]	Continuous	EC	0.30	*
Loamy sand [4]	Intermittent	EC	0.16	*
Loam-silty loam [5]	Continuous	EC	0.40	*
Keerange clay [6]	Continuous	Cl (NaCl)	0.55	80
Sandy loam[7]	Continuous	EC	0.50	74
Peat [8]	Continuous	EC	2.25	80
Greenfield loam [9]	Continuous	-	0.30 (1.5PV)	80
	Intermittent	-	0.68 (1.5PV)	80
Sandy loam [10]	Continuous	EC	0.40	80
Sandy loam[11]	Continuous	EC	0.37	80
	Intermittent	EC	0.37	80

Silty clay or clay [12]	Intermittent	% salt	0.50	80
Saline organic[13]	Continuous	EC	2.25	80
	Intermittent sprinkling	EC	2.25	80
Silty clay loam[14]	Continuous	EC	1.5	80
Yandera loam [15]	Continuous	EC	0.70	80
Sandy loam [16]	Saturated	EC	0.37	80
	Transient (Ponded)	EC	0.69	80
	Transient (Controlled application)	EC	0.47-0.52	80
Sandy loam [17]	Continuous	EC	0.60	80
	Intermittent	EC	0.60	80

PV = Pore volume, B = Boron, * not reported; Sources: [1]Beyce (1972), [2]Carter and Robbins (1972), [3]Chawla (1985), [4]Dahiya *et al.*(1981), [5]Dieleman (1963), [6]Garland (1966), [7]Gupta and Ramadurgaiaha (1984), [8]Hoffman (1980), [9]Jury *et al.*(1979), [10]Khosla *et al.*(1979), [11]Lefellar and Sharma (1977), [12]NEDECO (1968), [13]Prichard *et al.* (1985), [14]Reeve *et al.* (1955), [15]Talsma (1966), [16]Thokal (1990), [17]Verma and Gupta (1989)

Sprinkler Versus Continuous Leaching

Leaching efficiency in field soils could also be enhanced by controlling water application rate. Sprinkler irrigation system for this purpose could be a very handy tool to apply water at pre-determined rates. Oster *et al.* (1972) reported that intermittent ponding required 1/2 the amount of water that is required in continuous ponding while intermittent sprinkling required water that was intermediate of the two. Reassessment of the data from Hoffman (1980) also indicated that under similar conditions, sprinkler irrigation required more water than intermittent ponding at least when more than 60% salts were to be removed. One of the reasons for high water requirement in the case of leaching with sprinkler irrigation could be because more water is lost in evaporation as it takes more time to pass the same amount of water when applied at slower rate (Thokal, 1990). More water was required to leach the same amount of salt when water was applied at 1 cm/d compared to when it was applied at 2 or 3 cm/d (Fig. 3.10). In a recent study Prichard *et al.* (1985) did not find any superiority of leaching with intermittent sprinkling than with continuous ponding. In both cases 850 mm of water was required to remove 70% of salts present in 100 cm soil profile. This could be due to high moisture content prevailing in the sprinkler treatment and to low infiltration rate (3.5 mm/d) in the continuous ponding treatment after two weeks of ponding. In the Indian context, it appears that leaching by sprinkling may not play an important role unless sprinklers are purchased and used for irrigation purposes. Moreover, data to date suggest that similar conditions could be created through intermittent leaching. Therefore, usefulness of investment in sprinklers for leaching may be questionable.

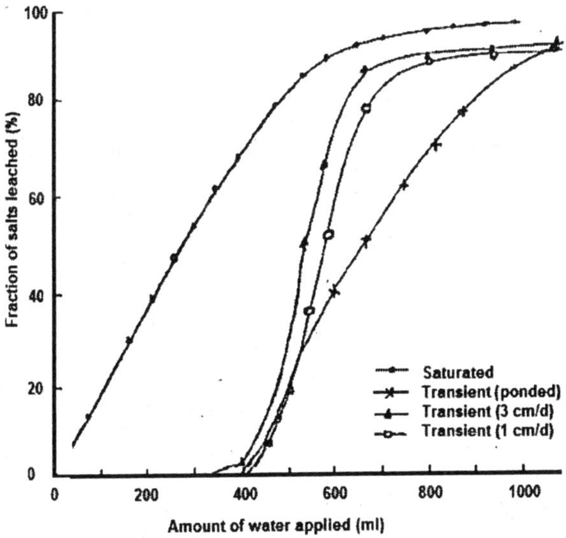

Fig. 3.10 Effect of water application rates on the fraction of salt leached

Spatial Variability of Soil Salinity and Adequacy of Leaching

One of the factors that affect the amount of water required for one time leaching is the initial salinity of the soil. Because of spatial variability in the soil physico- chemical properties; it is often difficult to assign a value to the initial salinity. The mean soil salinity value may not adequately represent the initial soil salinity for this purpose as in that case only about 50% of the land that has less salinity than the mean value will be reclaimed while the remaining 50% of the area expected to have higher initial salinity than the mean would remain unreclaimed. In a recent study on soil salinity appraisal, Thokal (1990) studied soil salinity variation in a saline land. Commonly used flood frequency analysis procedures were used to describe salinity distribution in a 0.25 ha area. The data revealed that normal distribution and cube root transformation failed to describe the data series. It was observed that for this site, EC_2

values at 50, 80 and 90% confidence level were 6.6, 9.7 and 11.6 dS/m. The calculated water requirement to bring the soil salinity at or below 1.0 dS/m was estimated to be 0.44, 0.52 and 0.56 cm/cm of the soil profile to be reclaimed (Fig. 3.11). It means to reclaim 80% and 90% area, the increase in water requirement is 18.2 and 27.3% respectively over and above the water needed to reclaim 50% of the area. It emerged that initially 80% area could be reclaimed as reclamation of 50% area might be too less. Moreover, 30% additional area could be reclaimed by applying only 18% more water. Amount of water required to reclaim additional area increases steeply beyond 80%. For example, to reclaim 90% area as compared to 80%, additional 7.7% water would be needed. The following guidelines should be useful while considering spatial variations in soil salinity.

- Take soil sample in adequate number from the area to be reclaimed.
- Fit a frequency distribution to the data series. For this purpose, log-normal, Gumbel, Pearson type III or one step power transformation could be tried.
- Calculate the initial salinity at 80% confidence level and use this estimate as the initial estimate of soil salinity.

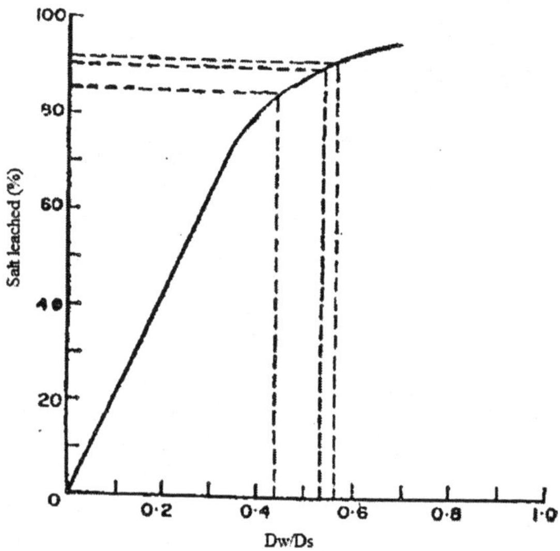

Fig. 3.11 Calculation of water required to leach a saline soil using a leaching curve

MODELING LEACHING

Empirical Equations

A number of empirical equations based on site specific investigations have been proposed as a practical approach to estimate water requirement for leaching (Table 3.10). Since soil parameters are not included in developing most of these equations, the equations are applicable only to situations where the soil, site and leaching conditions are similar to the experimental sites (Table 3.10). Few of them have been generalized and extensively tested (Dieleman, 1963; Hoffman, 1980). Reeve's (1957) equation is the most simple equation and has been used by other workers (Dahiya *et al.,* 1981; Lefellar and Sharma, 1977; Jury *et al.,* 1979). A critical analysis will reveal that equations proposed by Khosla *et al.* (1979) and Chawla (1985) are similar to the equations derived on the basis of layer wise equilibrium theory assuming the root zone as a single layer with $\theta=1/b$ and a=1. Similarly, the equation proposed by Raj and Nath (1984) also conforms to the single layer model of Burns (1976). When compared with this model, value of 'a' is 0.0 and b is equal to 1/2. Although other equations are not easily comparable, it appears that single layer models could be fitted with convenience to many field situations obviating the need to develop empirical equations. Khosla *et al.* (1979) also reported that Dieleman's and Burns models fitted as good to his data set (r=0.92 and r=0.91) as the empirical equation (r=0.96). In the equation reported in Table 3.10, the parameters D_l, D_d and D_s stand for depth of the water applied, depth of the water drained and depth of the soil, respectively. Parameters a and b are constants (Do not necessarily have the same meaning in different equations). f is the fraction of salts leached.

Layer Wise Equilibrium Theory

The approach based on layer wise equilibrium theory has been widely used to simulate leaching of saline soils. Currently used simple models are probably an offshoot of the chromatography approach. Basically the models are derived by dividing the soil into several layers and applied water into several pulses and adopting mass conservation in respect of both the solute and water. After each addition, the increment of water is assumed to completely mix with the soil solution in the top layer. A new equilibrium is assumed to occur between the adsorbed ion and that in the solution. If the field capacity is exceeded, the excess water is assumed to move into the next layer where it again mixes with the soil water in that layer. The solution in excess of field capacity is thus assumed to move into subsequent layers until either downward percolation ceases or the required depth is reached. Each successive pulse is treated in the same manner. Simple models based on treating the soil as a single layer and assuming constant moisture content could be treated analytically while other require listing of a computer program. Under both the situations mathematical treatment remains simple. These models could be described by an equation of the form given as eq. 3.22.

$$C_{i,j} = \frac{C_{i,j-1}\,\theta_{i,j-1}\,H_i + C_{i-1,j}\,(D_j - H_{i-1}\,(\theta_{i-1,j} - \theta_{i-1,j-1})}{\theta_{i,j-1}\,H_i + (D_j - H_{i-1}\,(\theta_{i-1,j} - \theta_{i-1,j-1})} \qquad (3.22)$$

It means, the salt concentration in the layer following j_{th} application of pulse of water each of some known depth will be equal to the concentration after the j-1_{th} pulse times the amount of water at the end of j-1_{th} application plus the concentration of the effluent from i-1_{th} layer times the depth of water drained below i-1_{th} layer divided by the total water. Here C_{ij} is the concentration of the i_{th} layer after j_{th} pulse, O $_{ij}$ is the moisture content of the i_{th} layer after j_{th} pulse, H_i is the depth of the i_{th} layer and D_j is the depth of water applied in j_{th} pulse.

Table 3.10 Commonly used empirical equations for calculating leaching

Equations	Proposed and used by	Parameter
$(D_f/D_s) = a.\exp(-b(1-f))$	Beyce (1972)	EC
$(D_f/D_s) = a - b(1-f)$	Beyce (1972)	Boron
$(1-f) = a.\exp(-b(D_d/D_s))$	Khosla *et al.* (1979),	EC, SAR
	Chawla (1985)	Chloride
$(1-f)(D_d/D_s) = a$	Hoffman (1980), Jury *et al.* (1979)	EC
$(1-f) = a + \exp(b.D_s/D_l)$	Raj and Nath (1984)	EC
$(1-f) = a\,(D_s/D_l) + b$	Reeve (1957), Lefellar and Sharma (1977)	EC
	Dahiya *et al.* (1981)	ESP
$\log \frac{(ESP - ESP_{eq})}{(ESP_0 - ESP_{eq})} = - a\,(D_d / D_s)$	Lefellar and Sharma (1977)	ESP
$D_l = a.\log(1/(1-f))^{-b}$	Agarwal *et al.* (1982) (Reported from Russian literature)	% salt content

In practice, within practical limits of accuracy, simple models based on the assumption that soil consists of a single layer describe the process of leaching. Gupta and Pandey (1983) and Thokal (1990) discussed the limitations of the existing models and pointed out that a low Peclet number is implied with the assumption of a single layer. A low Peclet number under field conditions is justified because of cracks, root holes and preferential flow paths. Although single layer models could be grouped into 4 classes depending upon the division of soil into sub-layers or of the applied water into pulses (Table 3.11), in practice, similar results are obtained from each group at least in the range of leaching required under field conditions.

Table 3.11 Grouping of the existing models on the basis of physical problem, number of pulses and number of soil layers

Group	Model proposed or referred as	Physical problem	No. of pulses	No. of layers	Prediction equations
A	Dieleman (1963)	1*	Infinite	1	1-exp (-P)
	Pandey and Gupta (1978)	2#	Infinite	1	(P+exp (-P) -1)/P
B	Burns (1975)	1	1	Finite	$1/(1+1/P.m)^{m/2}$
	Burns (1976)	2	1	Finite	$1/(1+1/P.m)^{m}$
C	Modified Burns (Gupta and Gupta, 1987)	1	1	Infinite	Exp (-1/2P)
	Modified Burns	2	1	Infinite	Exp (-1/P)
D	Thokal (1990)	1	Finite	1	$1-(1/(1+P/n))^{n}$
	Thokal (1990)	2	Finite	1	$1-1/P+1/P(1/(1+P/n))^{n}$

* uniformly distributed salts, # surface concentrated salts, m stands for number of sub-layers and n for number of pulses, P is pore volume drained.

van Der Molen (1956, 1973) contended that complete mixing in a large soil profile is difficult and therefore, it is necessary to sub-divide the soil profile into number of layers. Accordingly, a solution to estimate concentration of the water coming out of the last layer is given as

$$Cn = Co \exp(-t/T) \sum_{i=0}^{n-1} (t/T)^{i-1} \tag{3.23}$$

Here Cn is the concentration of the solute coming out at the effluent end of the nth layer, t/T is the residence time in a layer and i is the dummy variable. For leaching of surface concentrated solutes (salts present in the top layer only) the solution is written as

$$Cn = Co \exp(-t/T)\,(t/T)^{n-1} / (n-1)! \qquad (3.24)$$

It may be noted that t/T is equivalent to the pore volume of the layer.

It has been shown that these models give comparable results with the hydrodynamic dispersion theory provided Peclet number in the latter is taken twice that of the number of layers in a multilayer model. In other words, the Peclet number in the hydrodynamic dispersion theory is high when the number of layers in layer wise equilibrium theory are more. In practical terms, the dispersive movement of salts in the profile is less as the number of layers increase.

Leaching Efficiency

Leaching does not take place as a simple piston flow. If that was the case, one would need to apply only an amount equal to one pore volume to obtain complete leaching of salts present in the root zone. Several factors such as diffusion, dispersion and variation in localized flow velocities in the soil pores tend to reduce the effectiveness of leaching. Boumans and van Der Molen (1964) introduced the concept of leaching efficiency which says that a part of the water used for leaching bypasses the zone of reclamation without contributing to salt removal resulting in leaching efficiencies less than unity. It may be either due to incomplete mixing or due to movement of water through the large pores (cracks, root or rat hcles). The concept can be mathematically written as:

$$C_d = L_e C_s + (1-L_e) C_i \qquad (3.25)$$

It states that concentration of the drainage water, C_d, consists of a fraction, L_e, of the incoming water which will flow out of the root zone with the concentration C_s of the soil solution, and a fraction $(1-L_e)$ that will have its original concentration C_i. The part that has the same concentration as C_s i.e. L_e could be defined as leaching efficiency. Equation (3. 25) is then written as:

$$C_d = C_i + (C_s - C_i) L_e \qquad (3.26)$$

Miscible displacement theory as verified in field and laboratory trials indicates that leaching efficiency is well below 100% and it decreases with decreasing Peclet number as well as with the amount of water drained. Initial high leaching efficiency goes on decreasing as the reclamation proceeds. Thus, in practice leaching efficiency is a dynamic term and cannot be treated as a constant (Bouwer, 1969). As such, Bouwer (1969) assertion that leaching efficiency can be assigned numeric values on the basis of soil texture appears unreasonable (Schilfgaarde, 1976; Pandey and Gupta, 1981).

Leaching efficiency in addition to what has been stated above also depends upon the method of water application and initial distribution of salts in the profile. Leaching efficiency is also affected by the presence of cracks, root holes, animal burrows in the surface layer and continuous large pores in the profile as substantial amounts of water could bypass the zone of reclamation. It appears that in the original formulation of Boumans and van Der Molen (1964), only this part has been considered towards leaching efficiency which for many practical situations may depend upon soil texture. At this stage, it is necessary to distinguish between the leaching efficiency and the parameter, L_e. It is proposed that this parameter may be defined as bypass factor or mixing efficiency while leaching efficiency could be defined as the amount of salt drained as a fraction of the amount of salt present in the profile per unit pore volume after pre- decided pore volumes of water have been drained.

Rainfall and Soil Desalinization

Rainfall is of paramount importance in desalinization of saline soils or soils irrigated with saline water. Intensity and duration of the rainfall, duration of dry spells between two rain events and soil conditions including initial salinity and its distribution in the soil profile, leveling and bunding, condition of the top soil surface and moisture content of the soil profile at the time of rainfall, all influence the process of desalinization/leaching. In general, duration and degree of salt leaching or salt build-up over the seasons would greatly be dictated by the monsoon rains and the evapotranspiration. Monitoring of soil profiles before and after the monsoon season has revealed that leaching due to rainfall is substantial (Table 3.12). In general soils have low salinity after the monsoon or even at the time of sowing of next *rabi* crop provided saline water is not used to grow summer/ monsoon season crop (Fig. 3.12). Such regularities influence the way in which these soils are utilized.

Table 3.12 Change in soil salinity during monsoon season in representative soil profiles of arid and semiarid regions

Depth (cm)	EC_e (dS/m) Pre monsoon April/May	EC_e (dS/m) Post monsoon October	Depth (cm)	EC_e (dS/m) Pre monsoon April/May	EC_e (dS/m) Post monsoon October
Delhi Region[1] (Semiarid) 1100 mm rainfall (June-September)					
Ladpur (Nangloi)			**Raghopur (Najafgarh)**		
0-15	11.5	1.5	0-15	32.5	3.5
15-30	9.0	0.8	15.30	6.5	3.0
30-60	8.5	9.8	30-60	7.0	2.5
60-100	7.5	1.5	60-100	7.5	3.0
100-120	7.5	2.0	100-120	7.5	3.0
Paprawat (Najafgarh)			**Pahladpur (Alipur)**		
0-15	11.5	0.6	0-15	12.5	0.8
15-30	7.5	0.5	15.30	7.5	0.6
30-60	8.0	0.5	30-60	6.5	0.8
60-100	8.5	0.3	60-100	5.5	0.9
100-120	8.0	0.2	100-120	5.5	1.3
Jodhpur[2] (Semi-arid) 350-450 mm rainfall (July-September)					
Dantiwada (Jodhpur)			**Shikarpura (Jodhpur)**		
0-15	11.5	1.5	0-15	32.5	3.5
15-30	9.0	0.8	15.30	6.5	3.0
30-60	8.5	9.8	30-60	7.0	2.5
60-100	7.5	1.5	60-100	7.5	3.0
100-120	7.5	2.0	100-120	7.5	3.0

1 Paliwal and Yadav (1976), 2 Gupta and Abichandani (1970)

Kelley (1967) observed that if the precipitation is sufficient to penetrate to depths below the root zone, the accumulation of salts could largely be prevented. Bhumbla *et al.* (1964) reported that in light textured soils of regions with average annual rainfall equal to or more than 500 mm (monsoon rainfall 400 mm), highly saline water could be used to cultivate wheat crop as the soil salinity build-up during the winter season will be washed off during the monsoon season. Leaching through rainfall is significantly more effective than leaching by continuous ponding (Gupta and Abichandani, 1970). Studies on rainfall and soil leaching as described by Dhir (1977) indicate that fallowing is a useful technique to manage saline water in regions with low rainfall, in heavy textured soils or in years with below normal rainfall. Under these situations in western Rajasthan, farmers skip the winter cropping for a year so that additional leaching occurs during the second year and soils are fit to support the cultivation of winter crops.

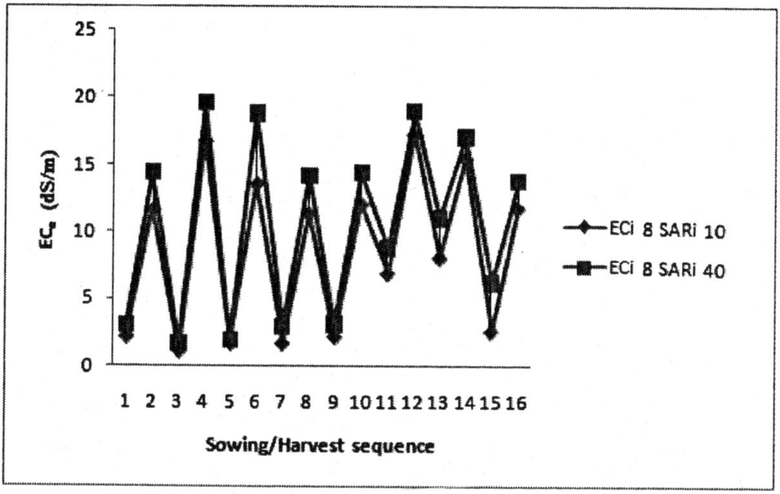

Fig. 3.12 Mean profile salinity at sowing and harvest of *rabi* crops at Agra

When Zuur-Verhoeven procedure was used to predict salinization-desalinization of saline water irrigated soils in India, it emerged that nearly 30% of the annual rainfall contributes to leaching in western Rajasthan. In these calculations, leaching efficiency of the applied water was assumed as 100%. Gupta and Pandey (1979) utilizing the same data concluded that at 3 out of 4 stations, contribution of rainfall towards leaching is 55%. The low utilization of rainfall at the fourth station (Jelwa) could be due to hard and compact nature of the soil. In these estimates single layer models were used to predict salinization-desalinization. These models account for the dynamic nature of the leaching efficiency which could vary from 100% initially to around 62% at one pore volume.

In a similar study on leaching due to rainfall in Haryana, Gupta (1985) showed that rainfall utilization in unbunded and unleveled fields at Sampla and Hansi was around 10% and 28% respectively. High run-off mainly contributed to low rainwater use efficiency at Sampla. The rainwater use efficiency increased to 70% and 55% upon bunding and leveling of the fields before the onset of the monsoon season. High efficiency at Sampla is ascribed to relatively high rainfall with relatively lesser duration and number of dry spells compared to Hansi.

Minhas and Gupta (1992) estimated the desalinization of soils irrigated with saline water due to rainfall under normal cultivation practices. The following equations were fitted to the heavy, medium and coarse textured soils respectively.

$$EC_a/EC_b = 0.1916\,(D_s/D_{rw}) + 0.0962 \qquad r = 0.75 \qquad (3.27)$$

$$= 0.1919\,(D_s/D_{rw}) - 0.0017 \qquad r = 0.95 \qquad (3.28)$$

$$= 0.1577\,(D_s/D_{rw}) + 0.0068 \qquad r = 0.66 \qquad (3.29)$$

Here, EC_a and EC_b are the soil salinity after and before the rainy season, dS/m; D_s is the depth of the soil, m and D_{rw} is the depth of the rainfall, m. The above relations reveal that about 1.85, 0.95 and 0.76 cm of rain water would be required to leach 80% of the salt from each cm of the soil needing salt removal. On the other hand, for medium textured soils about 0.5-0.6 cm of water per cm of soils would be required to leach about 80% of the salts. The larger amounts of rain water are needed to compensate for the runoff, rain water storage in the profile, evaporation of stored soil moisture in between two rain events and upward movement of salts during non-rainy days. Seasonal changes in the soil salinity for coastal saline soils revealed that in comparison to arid and semiarid regions, leaching due to rainfall is much less in the humid regions. High runoff due to heavy storms, poor rain water management practices and shallow water table contribute to lesser leaching of salts.

Conjunctive Use of Saline and Good Quality Waters

If good quality water is either limited or is unavailable at the desired time, conjunctive use of saline and good quality waters for leaching has immense potential. The following conditions should help in making efficient use of this technique (Reeve and Doering, 1966).

- The salinity and SAR of the saline water is less than the salinity and SAR of the soil
- Gradually low EC and SAR water is used for leaching followed by fresh water in the final stage of leaching. Conjunctive use requires that last stage be synchronized with availability of good quality water.

A number of studies have validated the application of this concept. Attempts have also been made to quantify the saving in fresh water with conjunctive use in comparison to when fresh water alone is used (Gupta and Pandey, 1983; Ranade, 1984; Chawla, 1985). Gupta and Pandey (1983) observed that although 18-33% more water (saline + fresh) is needed in conjunctive use, the fresh water requirement is reduced by 15–40% depending upon the salinity of the ground water, initial salinity of the soil and fraction of the salts drained with saline water (Table 3.13). Chawla (1985) observed that in sandy loam soils saving of fresh water is in the range of 48.3 to 61.2% although the total water requirement is increased by 37.9 to 63.8%. Typical leaching behavior of a saline soil with conjunctive and fresh water alone is shown in Fig. 3.13.

Table 3.13 A typical calculation to estimate saving in fresh water requirement

Saline soil = 100 dS/m

Leaching water = 40, 20, 0 dS/m

Final salinity desired = 4 dS/m

Fraction of the salt retained=0.04

EC of leaching water	$(C_s$-$C_i)/C_s$* (%)	Percent of salt leached	Saline water required (Number of pore volumes)	Fresh water required (Number of pore volumes)	Total water required (Number of pore volumes)	Increase in water required over A (%)	Saving in fresh water required over A (%)
40	0.6	48	1.59	2.70	4.29	34.1	15.6
		54	2.30	2.46	2.46	48.7	23.1
20	0.8	64	1.59	2.20	2.20	18.4	31.3
		72	2.30	1.95	1.95	32.8	39.1
0	1.0	96	3.20		3.20 (A)	-	-

*C_s is the soil salinity and C_i is the salinity of leaching water

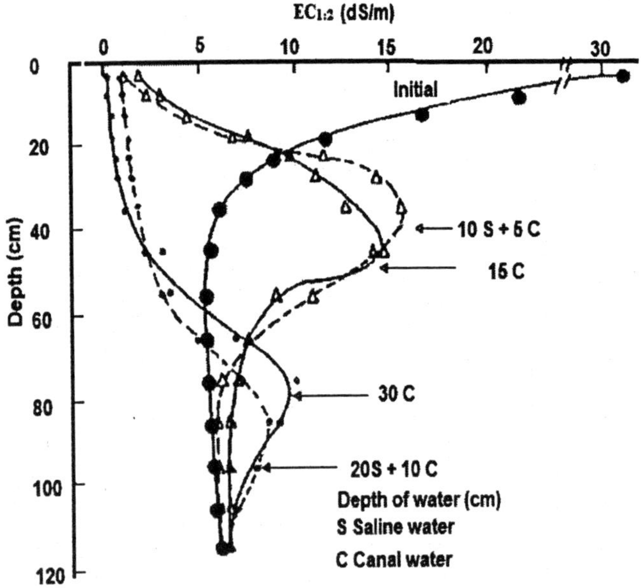

Fig. 3.13 Leaching behavior of saline soils with conjunctive use of saline and fresh water

LEACHING FOR SALT BALANCE

Salts are added to the soil along with the application of irrigation water. Sustainability of irrigated agriculture demands that amount of salts added must not be allowed to accumulate beyond a certain limit. Surplus salts must be removed by the drainage water leaving the root zone. If mass of salts added from weathering of minerals or dissolution of salt deposits, mass of salt added as fertilizers or mass of salts lost as precipitation and mass of salt taken up by crops are neglected and it is assumed that under steady state conditions, there is no build up of salinity, then for the mass balance of salts in the root zone, we have:

$$D_i C_i = D_d C_d \tag{3.30}$$

Where, D_i and D_d stand for the amounts of irrigation and drainage waters and C_i and C_d are the salt concentrations of these waters. From eq. (3.30), it immediately follows that

$$\frac{D_d}{D_i} = \frac{C_i}{C_d} \tag{3.31}$$

The leaching requirement, LR, defined as the fraction of the irrigation water that must pass through the root zone to maintain soil salinity at some pre-decided level could be mathematically written as:

$$LR = D_d/D_i \text{ or } C_i/C_d \quad (3.32)$$

As electrolyte conductivity (EC) of the solution is a good index of the solution concentration, solution concentration is often replaced by EC such that eq. (3.32) is written as:

$$LR = EC_i/EC_d \quad (3.33)$$

Here, EC_i and EC_d are the electrolyte conductivities of the irrigation and drainage water respectively. Recent literature makes distinction between LR which is the theoretically average requirement for leaching and LF which is the actual fraction of the applied water that passes through the plant root zone (Smith and Hancoc, 1986). Similarly distinction is also made between LR calculated from solution concentration as in eq. (3.32) and LR calculated from solution EC as in eq. (3.33). Smith and Hancock (1986) have shown that universal substitution of electrolyte conductivity for salt concentration is objectionable as LR estimated from C_i/C_d or EC_i/EC_d may not be equivalent.

Estimating Leaching Requirement for Salt Balance

To estimate LR, one must know C_d or EC_d. For a good approximation of this parameter, Bernstein (1961) suggested the use of average EC_e at which 50% yield reduction occurs as the value of EC_d in case of forage and vegetable crops. For fruit crops, EC_e at which 10% yield reduction occurs should be taken as EC_d. Bernstein and Francois (1972) postulated that yields are nearly insensitive to the salinity of the lower root zone and therefore, the EC_e values at 50% or 10% reduction in yield could be increased by 4 times before use in eqs. (3.32 or 3.33). Schilfgaarde *et al.* (1974) while commenting upon several equations notably the one proposed by Bernstein (1964) suggested that EC_d could be assumed twice of the EC_e at which yield reduction is 100%. A more logical semi empirical model of Rhoades (1974) states that average salt concentration in the profile could be estimated from the salinity at the top and bottom of the root zone such that:

$$C_S = K\,(C_t + C_l)/2 \quad (3.34)$$

Here, C_l is the salt concentration at the bottom and C_t at the top of the root zone. K is a constant and its value is governed by the shape of the salt profile in the root zone. Equating C_S to the average root zone salinity for a pre-determined yield reduction, LR in terms of electrolyte conductivity can be written as:

$$LR = (EC_i/(2EC_s/K - EC_i) \quad (3.35)$$

This equation has been used by Ayers and Westcott (1976) assuming K=0.8 and EC_s= 2 EC_e. The proposed eq. (3.35) overestimates LR except at low leaching requirements. Since K is a weighing factor to account for salt distribution pattern in the profile, Gupta and Yadav (1988) suggested that K in eq. (3.35) could also be a function of LR. It was proposed that K should be calculated by the following relation:

$$K = 0.7 + 0.30\,LR \quad (3.36)$$

With this change eq. (3.35) can be written as:

$$EC_i/EC_s=2\,LR/(1+LR)(0.70+0.30LR) \quad (3.37)$$

It emerged that modified Rhoades equation (eq. 3.37) gave better estimates than the original and to a greater extent reduced over estimation at high LR. Hoffman and van Genuchten (1983) also proposed few equations assuming water uptake to be exponential, trapezoidal or in the ratio of 40:30:20:10 from the four layers of root zone, each layer being 25% of the total root zone depth (Fig. 3.5). They recommended that linearly averaged root zone salt concentration equation based on exponential uptake function should be used to determine LR. The LR estimated with the help of this equation is close to the LF obtained under field experiments. The equation in a general form and in terms of EC, could be written as:

$$EC_s/EC_i=1/LR +(H/Z\,LR)\ln(LR +(1-LR)\exp(-Z/H)) \quad (3.38)$$

Here Z is the depth of the root zone and H is an empirical constant. The best results are obtained when H is set equal to 0.2 Z. Smith and Hancock (1988) proposed an equation essentially similar to the one proposed by Hoffman and Van Genuchten (1983). It could be used to estimate the uptake weighted average root zone salt concentration. The equation is written as:

$$EC_s/EC_i= \ln (LR)/(LR-1) \quad (3.39)$$

A comparison of four equation namely (3.35), (3.37), (3.38) and (3.39) revealed that Hoffman and Van Genuchten (1983) equation has an edge over other equations. Based on eq. (3.38), a nomograph to calculate leaching requirement for some combinations of soil salinity or the salinity of the irrigation water or the crop threshold is presented in Fig. 3.14.

Appropriate Time of Leaching

The time of leaching is not critical as long as soil salinity remains within crop tolerance limit during crop growth as well as during critical growth stages. Otherwise one has to apply extra water to leach down the salts. In most cases, monsoon rainfall should be adequate to accomplish leaching. If necessary and conditions permit, annual leaching during no cropping season should be preferred. If excessively high saline water is used for irrigation, leaching at each or after few irrigations or after a season might be required. Considerable saving of water is made when periodic leaching is practiced than applying additional water with each irrigation (Arar, 1971). For soils with low infiltration rates, it is better to postpone leaching until after cropping so as to reduce the time of water stagnation and consequently loss in crop yield.

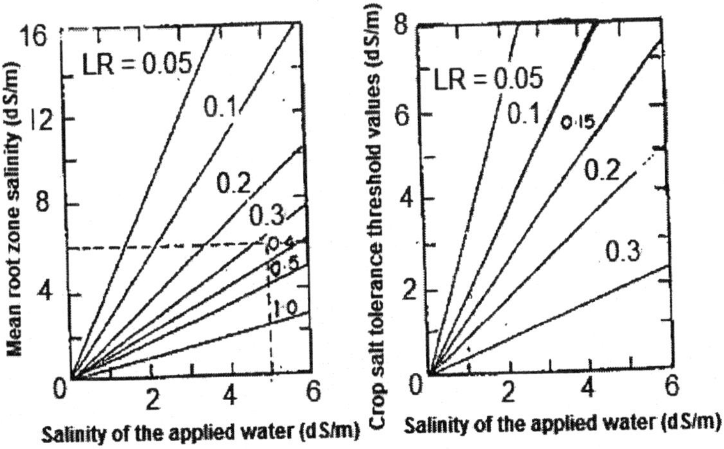

Fig. 3.14 Nomograph for eq (3.38) to estimate leaching (left) with mean root zone salinity and (right) with crop salt threshold

Leaching Requirement under Field Conditions in India

Leaching requirement concept has been proposed assuming that steady state conditions would develop upon long-term irrigations. Due to monsoon climatic conditions, steady state conditions are not established in irrigated soils of India. Therefore, under the salinization- desalinization cycles experienced during a year, validity of the LR concept is doubtful. Wherever applicable, leaching requirement of crops varies in a range of 3 to 17%. Hoffman and van Genuchten (1983) reported that LR could be as high as 29% as observed for alfalfa irrigated with saline water of 4 dS/m. On the other hand, the average water application efficiencies of five irrigation projects in India were observed to be 40% in one, 50% in three and 70% in one (Bos, 1977). Since field water application efficiencies indicates the quantum of deep percolation losses, it could be visualized that a minimum LF of 0.3 is obtained under field conditions if one assumes that water is uniformly distributed. Assuming a fair water distribution efficiency and monsoon rainfall in arid and semiarid regions, annual leaching fraction would be sufficient to meet LR requirement of most crops. With present practices, there is little chance of improving application efficiencies beyond 70% under surface irrigation. It is doubtful, therefore, that there is any significance of the leaching requirement concept under the current irrigation management systems. As a testimony, field results of Jain *et al.*(1976) indicate that yield of wheat under irrigation with saline-sodic water (EC = 4.8 dS/m, RSC = 7.4 meq/L, SAR = 32) did not increase by application of 25% extra water. Poonia *et al.* (1974) also observed that levels of leaching (0, 25 and 50%) did not have much effect on wheat and pearl millet grown with saline water.

In spite of the limitations, future developments such as increasing use of sprinkler and drip irrigation and expansion of protected cultivation in the country, the LR concept may provide useful guidelines in the use of saline water. Even in surface irrigation systems, when distribution efficiency is poor, leaching fraction may play an important role in controlling salinity build-up in soils.

Cropping during Leaching

In most barren saline soils, soil salinity is too high to grow any crop. It usually takes two to four years to fully reclaim a soil. It is due to periodic migration and accumulation of salts within the profile or transport of salts into the root zone through capillary rise from the shallow ground water table. Suitable crops need to be identified for cultivation during this intervening period before the lands are fully reclaimed. This state is usually referred to as the leaching/cropping stage. The crops during this period should be selected on the basis of salt tolerance, tolerance to water logging, water requirement-water supply scenario and economic value. A commonly made recommendation that salt tolerant crops should be grown during initial years should be viewed carefully. For example; if barley, a salt tolerant crop is cultivated, there is likelihood for the salinity level in the surface layer to increase compared to a crop like wheat or *berseem* (clover) owing to its low water requirement. Similarly, rice crop is often considered inappropriate because it is less tolerant to salts. On the other hand, if water could be arranged for a season, salt level in the profile could be largely reduced to allow a greater choice of crops in the forthcoming season. It has now emerged that one rice crop could be included in a long-term crop rotation on such lands even after reclamation.

The cropping of the barren lands could begin either from the monsoon season or from the winter season depending upon how much reclamation is achieved before the optimum sowing period. Dahiya *et al.* (1981) transplanted 6 week old rice seedlings just after submergence and without any previous leaching except from rainfall. A good rice crop was obtained in all the plots (with or without previous leaching) because soon after submergence, salinity in the top 10 cm layer reduced to a level that 6 weeks old seedling were able to tolerate that level. During the cultivation of rice, 223 mm of water passed through 100 cm soil depth and therefore, good wheat and *Sesbania* crops could be taken in the following season. The yield differences between intermittent and continuous ponding were observed to be non significant because the salinity in the two treatments dropped well below the tolerance limits of the crops grown. On the other hand, when reclamation began with *Sesbania* or wheat, the amount of water applied during pre-sowing leaching had a significant effect on the crop yield (Table 3.14). In wheat, although no significant differences were observed beyond 43.5 cm yet numerically yield increased until 63.5 cm of water applied.

Table 3.14 Effect of application of leaching water on crop yield

Leaching water (cm)	*Sesbania* green manure (t/ha)	Total leaching water (cm)	Wheat (t/ha)
0	33.9	33.5	1.8
10	86.3	43.5	27.4
30	113.1	63.5	33.5
50	119.5	83.5	32.1
70	117.2	103.5	35.0
CD at P = 0.05	8.62		

Source : Dahiya *et al.* (1980)

Gupta (1981) studied the effect of leaching on pearl millet and wheat crops grown in sequence. About 43.2 cm of rain water was conserved in bunded and leveled fields and additional 17.0 cm water was applied to irrigate the pearl millet crop. The pearl millet yield was very poor when grown without pre leaching compared to pre leaching with 20.0 cm of water. The yield improved from 0.3 t/ha to 1.25 t/ha (Table 3.15). The overall yield was poor because initial salinity was high and drainage was not provided which lead to the resalinization of the plots during the crop growing season. Leaching carried out during the monsoon season helped better growth of wheat following pearl millet crop (Table 3.15). Gupta (1982) also attempted to reclaim these soils beginning with *rabi* season, wheat being the first crop. Field plots were bunded and leveled before the onset of the monsoon season. As much as 80 to 90% of the initial salts present in the top 60 cm profile were leached during the monsoon season. However, initial sampling showed wide variation in soil salinity allowing the division of plots into three categories namely low, medium and high salinity. The amount of water applied as pre leaching treatment resulted in significant differences in yield only in case of medium and high salinity plots. Similar results were also obtained in plots where barley was grown as the test crop. As the threshold soil salinity for wheat and barley is estimated at 4.5 and 6.8 dS/m respectively, significant differences were not observed in the plots with low salinity. The minimum amount of water applied lowered the soil salinity at or below the threshold level.

Table 3.15 Effect of differential water application during summer on wheat yield during winter

Water applied during monsoon (cm)		Yield(t/ha)	
Rainfall	**Irrigation**	**Pearl millet**	**Wheat**
32.0	0	-	1.2
43.2	0	-	3.2
43.2	17	0.30	3.8
43.2	20*+17	1.25	4.5

Source : Gupta (1982)
* water applied to leach the salts before sowing

Practices to Enhance Leaching Efficiency

Leaching is the only practical means to reduce soil salinity and bring saline soils under the plow. To reclaim the soils with limited amount of water or to achieve early reclamation, a number of practices can be used. Some useful practices are summarized in Table 3.16.

Table 3.16 Practices to modify leaching behavior

Practice	Effect
Sub-soiling	• Improves infiltration rate • Removes/reduces natural channels
Tillage with inversion	• Improves infiltration rate of the soil • Physically moves the surface layer with high concentration of salts to lower depths
Chemical amendment	• Improves infiltration rate • Controls surface crusting • Maintains soil structure
Breaking hard pan (Auger hole piercing)	• Improves water movement through the hard pan • Reduces chances of water table build-up
Irrigation practices	• Improve leaching efficiency by maintaining high moisture content • Situations resembling intermittent leaching could be created

Pitfalls

Quite often, leaching is practiced as a simple application of excess water before a crop is sown. In practice, such a simple interpretation might result in delays or even failure of a reclamation program. Practically speaking, leaching is beneficial only when applied as a package. One or more of the following pitfalls are often noticed in the design of leaching programs.

- Leaching without adequate drainage often leads to severe problem of water logging and soil salinization. Right type of drainage system with appropriate capacity must be installed at the right depth before starting a reclamation program including leaching (see Chapter 2).
- Inadequate bunding, leveling, field layout and slopes often result in spotted leaching.
- Arrangements for water in right quantities and at the right time should be ensured. Inadequate application resulting in partial leaching may lead to failure of the whole program.
- If delay is feared due to low permeability of soils, then cultural practices which enhance infiltration rate and/or suitable

amendments should be applied at the right moment. Many a times this aspect is completely ignored.

- Steps should be taken to minimize secondary salinization, which in most cases is the principal mechanism leading to delay/failure of a leaching program.

4

RECLAMATION OF ALKALI LANDS

Reclamation and management of alkali soils poses serious problems due to their poor physical and chemical properties. These soils have excess concentration of sodium ions (Na^+) in the soil solution phase as well as on the cation exchange complex and as such exhibit unique structural problems (slaking, swelling, and dispersion of clay) and specific conditions (surfacing, crusting and hard setting). These conditions seriously affect water and air movement, plant available water holding capacity, root penetration, seedling emergence, runoff, erosion, tillage and sowing operations. High pH due to dominance of CO_3^{2-} and HCO_3^- of sodium and high ESP adversely affect crop productivity. Because of the high pH and high sodium, terms alkali soil or sodic soils are synonymously used to describe these soils. This chapter discusses some of the interventions required to reclaim and manage alkali lands.

CHARACTERISTICS OF ALKALI SOILS

- Characterized by high pH_s (> 8.2) and high ESP (> 15)
- Occur in regions with flat topography with slope varying from 0.1 to 0.2% with poor drainage conditions
- Low infiltration rate, < 5% of the rate in normal lands in the vicinity
- Black patches may be observed at the soil surface because of the dissolution of organic matter at high pH, which is transported to the soil surface through capillary action. It is why these soils are sometimes referred as black alkali soils
- Presence of *Kankar* layer ($CaCO_3$ concretions) at depth ranging from 0.5 m to 1.8 m depth
- Commonly encountered in semiarid and sub-humid agro-climatic conditions with annual rainfall of more than 500 mm.

High intensity storms of 15-20 cm depth occur in one day and monsoon season might have a long dry spell(s) of up to 30 days

- Underlain by good or alkali ground water

FACTORS AFFECTING PLANT GROWTH IN ALKALI SOILS

Plant growth, yield and/or quality of crops in alkali soils are affected by one or a combination of the following factors, the extent depending upon the amount of exchangeable sodium, pH, and nature and kind of crops grown, macro and micro-environmental conditions and management interventions (Gupta *et al.*, 2007).

Chemical Factors

- High pH due to high concentrations of HCO_3 and CO_3
- Toxicity of Na, HCO_3 and CO_3 ions
- Excess of Na causing imbalance in the uptake of other ions due to antagonistic effect such as K and Ca nutrition. Deficiency of Ca is encountered since nearly all the soluble Ca is precipitated as insoluble $CaCO_3$
- Decreased solubility and availability of micronutrients like Zn, Fe and Mn due to high pH and presence of $CaCO_3$ and soluble CO_3 and HCO_3 ions
- Increased solubility and accumulation of toxic elements like F, Se and Mo in plants that may affect crop yield and/or health of the animals feeding on such crops
- Presence of hard $CaCO_3$ layer, which act as a physical barrier for the movement of water, salts and roots of crops and trees.

Physical Factors

- Poor water and air permeability as a result of high dispersion of soil aggregates and clay particles
- Low availability of water due to poor conductance from the lower to the upper soil layers and restricted effective root zone
- Hard crust on the surface layer which greatly hinders the seedling emergence and reduces the germination percentage and thus plant population

- Poor drainage resulting in aeration problems
- Water stress during long dry spells
- Difficulties in preparing seedbed as all parcels of the land do not attain proper moisture conditions at the same time; this also results in clod formation and poor tilth of the soils.
- Low activity of useful microbes due to high pH and excess exchangeable Na
- Continuous loss of fertile soil due to air and water erosion
- Low organic matter content.

THE RECLAMATION PACKAGE

The reclamation package released by CSSRI, Karnal consists of the following major components.

- Land leveling and construction of *bunds* (field dykes)
- Assured source of good quality irrigation water
- Improved surface and subsurface drainage
- Application of amendments
- Leaching
- Selection of crops and cropping sequences
- Nutrient management
- Improved cultural practices

A brief description of each component of this package is described in the following sections.

LAND LEVELING AND CONSTRUCTION OF *BUNDS*

Any reclamation program begins with appropriate land development. The first activity in this context is to dyke the fields (construction of bunds). This activity is essentially needed for the two reasons. Firstly, the rainwater stored in the field after gypsum application should not go out of the fields. Secondly, runoff from the surrounding unreclaimed areas should not enter the fields. As such, construction of 45 to 60 cm high dykes is crucial for an efficient reclamation program. During the first few years, these soils are prone to erosion by the flowing water and therefore, slightly wide and relatively high dykes need to be constructed (Fig. 4.1).

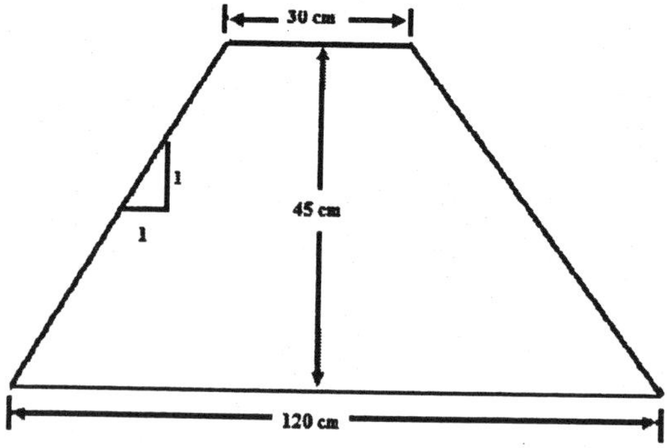

Fig. 4.1 Cross-section of the bund at the time of initiation of land reclamation

After the construction of bunds, fields are properly leveled. Even small differences in the level would result in non-uniform leaching causing patchy crop growth. However, the term leveling should not be misunderstood as tabletop leveling. Instead it should be treated as grading in the direction of irrigation. Precise grading rather than tabletop leveling is essential to carry out efficient irrigation and surface drainage. As far as possible, field should be provided with a slope of 0.1% to drain excess water; the range being 0.05 to 0.20%. It is estimated that about 10 tractor h/ha would be enough to accomplish the land smoothening work. Studies have shown that a super leveler could effectively complete the task with sufficient accuracy (Table 4.1). Presently, laser land leveling is increasingly used which results in a more precise surface and grade (Table 4.2 and Fig. 4.2). The data reveal that the range within which the ground levels deviate from the mean improves with land leveling. The improvement is much superior in the laser than conventional land leveling. The leveling index (LI) with laser leveling could be brought to less than 1 while it remained more than 2.0 in conventional leveling (Table 4.2). The best time to undertake *bunding* and leveling operations is April-May. It helps to finish all the activities before the first crop of rice usually transplanted in the month of July.

A well- planned field layout should then be prepared to ensure that each field receives water and is connected to collector and/or main field drain and is fully trafficable. The irrigation system should deliver water uniformly to all parts of the field, while the drainage channels are capable to remove excess irrigation or rainfall (See Chapter 2). In border irrigation, a drainage channel may run perpendicular to the borders at the tail end. The estimated length of irrigation and drainage channels for one ha of land is about 100 m each.

Table 4.1 Time, cost and land uniformity coefficient (LUC) for different leveling implements

Implement	Time of operation (h)		Time for land shaping operation (h)	LUC
	Plowing	Discing		
Leveling blade	10	10	26.4	0.72
Leveling float	10	10	15.0	0.63
Leveling blade + float	10	10	18.5	0.90
Super leveler	10	12	19.5	0.96

Source: Tyagi and Singh (1979)

Table 4.2 performance parameters in laser and conventional land leveling

Field	Mean value	Range		LI	
		Before	After	Before	After
Laser 1	2.102	2.055-2.155	2.074-2.124	1.56	0.80
Laser 2	2.860	2.790-2.940	2.830-2.890	2.29	0.90
Conventional	2.860	2.784-2.980	2.817-2.917	3.00	2.10

Fig. 4.2 A view of laser land leveling operation in the field

ASSURED SOURCE OF IRRIGATION WATER

Timely application of water in adequate quantities is essential for the successful implementation of a reclamation program. In general, alkali areas are underlain by good quality ground water and the aquifer characteristics are good enough to install a tube well. Since, canal water is limited and undependable; it is advisable to install a tube well before embarking upon a reclamation program. Depending upon the aquifer characteristics, a cavity or a strainer type tube well could be installed. A tube well of 10 L/s capacity is sufficient to cover an area of 4 ha. A cavity well serves as a dual purpose structure. It provides water for irrigation and helps to lower the water table as land reclamation work progresses (Fig. 4.3). Currently, when water table is declining at a very fast rate; a submersible pump might have to be installed. It would naturally impact the cost of land reclamation.

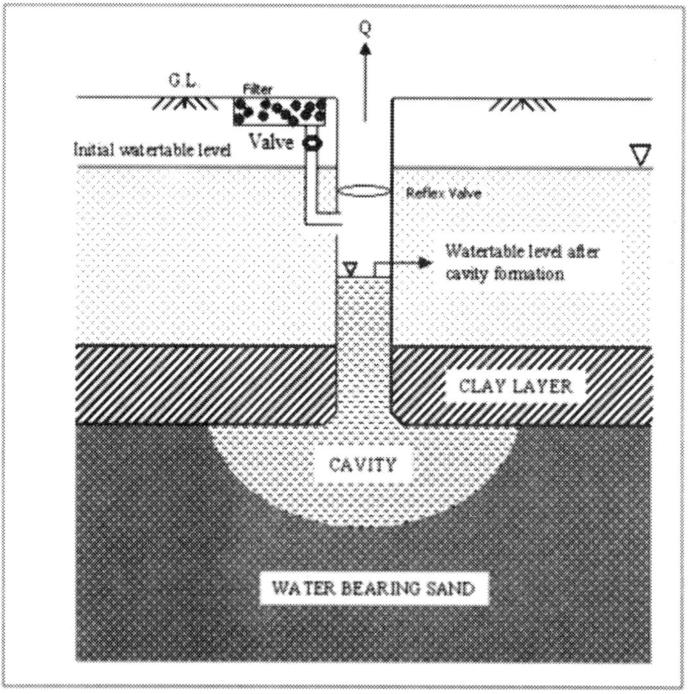

Fig. 4.3 A multi-purpose cavity well for irrigation, drainage and artificial ground water recharge

Rainwater management could also play a very significant role in conserving good quality water to reduce dependence on ground water for irrigation. Three tier rainwater management technology developed at CSSRI, Karnal can be implemented as it will also ensure timely and adequate surface drainage. The technology consists of the following three components:

- Rainwater is allowed to accumulate and remain on the crop land till such time and extent that it does not adversely affect the crop. It significantly reduces the water required to cultivate rice and also minimizes the drainage requirement.
- Excess water from the fields is led to the dugout ponds or natural depressions of sufficient capacity (Fig. 4.4). The stored water is reused during dry spells or for irrigating winter season crops. Later on it could also be developed as a multi-enterprise farm with aquaculture. It should help improve the livelihood sources and provide high and regular income to the farmers.
- Rainwater in excess of these two compartments of storage is led out of the area to the natural drainage system through tertiary and main drains or is used for ground water recharge.

Fig. 4.4 A view of the rainwater stored in on-farm pond, the second tier of the technology

SURFACE AND SUBSURFACE DRAINAGE

Since alkali soils are characterized by low infiltration rate, drainage is essential to ensure high productivity. The adverse effects of water logging in these soils appear due to a combination of poor aeration, reduced uptake of plant nutrients, imbalance between the uptake of various ions and toxicity effects of various elements/cations. Temporary water logging situation is encountered during monsoon season or following a low intensity shower especially occurring after a normal irrigation during the winter season. The adverse conditions continue to prevail even few days after the water has been removed from the soil surface. Even the reclaimed alkali lands require efficient and effective surface drainage system as revealed by yield losses from short-term water stagnation (See Table 2.1). It may be noted that except for the barley, yield reduction exceeds 10% with water stagnation of 2 days in all the test crops (Table 2.1, Gupta *et al.*, 2004). Therefore, in semiarid and sub-humid regions where alkali soils occur in abundance, improved surface drainage is an essential element of a reclamation plan. The three tier system of rainwater management discussed in the previous section is a good intervention to ensure adequate and timely surface drainage especially in land locked areas.

In areas where water table exists at shallow depth and especially where ground water quality is poor, option of installing some kind of subsurface drainage needs to be explored. It has been established that areas underlain with fresh ground water are amenable to vertical drainage (tube well drainage). The alkali areas in Haryana, Punjab and Western Uttar Pradesh are typical examples where shallow water table has been managed with vertical drainage. The tube wells in these areas acted both as a source of irrigation and drainage. Although, vertical drainage has proved to be the most effective at these places yet at places where rainfall is high, canal water is available in relatively more quantity, ground water is alkali in nature and resource endowment of the farmers is relatively poor, it has not been adapted to the desired extent. Conventional pipe drainage to lower the water table may pose problems because of low hydraulic conductivity of these soils. As such, an integrated solution consisting of surface, pipe, vertical and bio-drainage needs to be adopted to hasten the land reclamation process. These drainage technologies have been discussed in Chapter 2 of this book and readers may refer to finalize an appropriate drainage plan.

APPLICATION OF AMENDMENTS

To have successful cropping in alkali lands, pH/ESP of the soil must be brought down to acceptable level, for which chemical amendments are primarily used. These amendments neutralize the pH of the soil, react with some of the free sodium carbonate, and replace the exchangeable sodium by calcium. In order to minimize the reaction of the amendment with sodium carbonate and bicarbonates in the soil solution, a heavy pre-sowing irrigation is usually given before the amendment is applied. The degree of soil alkali (ESP) and the availability of *Kankar* (impure $CaCO_3$) in the soil help to decide the kind of amendment to be applied as well as determine the effectiveness of the chemical amendments.

Classification of Chemical Amendments

The chemical amendments are broadly classified into three categories:

- Soluble calcium salts, viz. calcium chloride, gypsum and phospho-gypsum which directly supply soluble Ca for the replacement of exchangeable Na
- Acidifying materials, viz. pyrites, sulphuric acid, and industrial wastes (spent wash, dairy waste) which due to their acidulating action and/or increasing the partial pressure of CO_2 supply Ca indirectly by dissolving calcite natively found in alkali soils. The Ca so mobilized is used to replace Na from the exchange complex
- Organic materials viz. rice husk, FYM and green manure etc., which helps to improve soil structure as well as dissolve the native $CaCO_3$ albeit weakly. The reclamation efficiency is usually low with these kinds of amendments.

A major factor that determines the use of a particular amendment is its cost. Mined gypsum being cheaper has been the most preferred amendment for the reclamation of alkali lands.

Calcium Sulphate

Three main commercial forms of calcium sulphate are: Anhydrite form - land plaster ($CaSO_4$); semi-hydrate form- commonly called plaster of Paris ($CaCO_4.1/2H_2O$) and di-hydrate form- generally known as gypsum ($CaSO_4.2H_2O$). Di-hydrate kind of gypsum is commonly used in agriculture.

The chemical amendment, $CaSO_4.2H_2O$ reacts with exchangeable sodium to replace it with calcium on the exchange complex. It directly supplies Ca to replace the exchangeable Na as per the following reaction:

$$2\ \text{Na-soil} + CaSO_4.2H_2O \longrightarrow \text{Ca-soil} + Na_2SO_4 + 2H_2O \qquad (4.1)$$

Alkali soil — Normal soil

Pyrites

A major supplier of the pyrites in India, Ms. Pyrites Phosphates and Chemicals Limited, mines the pyrite mineral along with pyritepherous shells from mines in Sikar district of Rajasthan and Amjhor district of Bihar. Pyritepherous shells besides containing iron have sulphur, which acts as an amendment to reclaim alkali soils. The low-grade pyrites, containing about 22% sulphur, has been used as a soil amendment. Pyrites (FeS_2) furnishes Ca indirectly, by first oxidizing to an acid, which in turn reacts with soil lime to furnish soluble calcium.

$$2FeS_2 + 2HOH + 7O_2 \longrightarrow 2FeSO_4 + 2H_2SO_4 \qquad (4.2)$$

$$CaCO_3 + 2H_2SO_4 \longrightarrow CaSO_4 + CO_2 + H_2O \qquad (4.3)$$

$$2\ \text{Na–soil} + CaSO_4 \longrightarrow \text{Ca–soil} + Na_2SO_4 \downarrow \qquad (4.4)$$

The oxidation of pyrites begins as soon as the material is mined and exposed to the air. The reaction is slow yet it is wasteful as some of the S may escape to atmosphere as follows:

$$FeS_2 + 3O_2 \longrightarrow FeSO_4 + SO_2 \uparrow \qquad (4.5)$$

The oxidation process is mediated through the bacteria *Thiobacllus ferrooxidens*, which requires a warm, well aerated and moist soil and a pH of 3 to 4 for its efficient working. However, when pyrites is added to the alkali soil, the acid already contained in it reacts with $CaCO_3$ of the soil to release Ca for exchange with Na on the clay complex. As such, pH in the vicinity of pyrites granules increases adversely affecting the biological oxidation of pyrites. Thus, oxidation of S in pyrites is incomplete and its efficiency remains below that of gypsum when applied on equivalent S basis. Many studies have shown that its efficiency is about 70 to 80% as compared to gypsum at the recommended dose of application. Later studies revealed that the soluble S content of the material at the time of field application rather than the total S content, determines its reclamation efficiency. Freshly mined pyrites contains many times less soluble S than the pyrites stored under a rain shelter for few years. Depending upon storage conditions, the soluble S content of pyrites varies from 1 to 11%. As a thumb rule, for pyrites to be effective, it must contain at least 5 to 6% soluble S at the time of application. Reclamation efficiency of the pyrites could be increased as follows:

- The pyrites must be crushed properly before use. Its soluble S content should be more than 5%
- Apply pyrites only when the field moisture content is around field capacity
- For proper oxidation, it should be applied as thinly as possible. Besides, sufficient time should be allowed to oxidize the material before the leaching process is initiated
- The soils should be leached with good quality water for 7-15 days preferably in intermittent leaching mode before translanting the first rice crop.

Pyrites containing 6-8% soluble sulphur has a pH of 2-2.5 when dissolved in water (1:100). As such, it is used as a field test to assess the quality of the pyrites for land reclamation. The test procedure is as follows:

Dissolve 1-g pyrite in 100 ml distilled water and check the pH. If the pH is more than 2.5, the pyrite must be stored periodically moistening by sprinkling water over the bags. The material should be stored until the test reveals a pH of 2.5 or less. One may use a litmus paper or a pH meter for this test.

Precaution

Since pyrites contains free H_2SO_4, proper care must be exercised during field application. Gloves and gumboots should be used to avoid acid injury.

Sulphur and Sulphuric Acid

Alike pyrites, mineral sulphur is also converted to H_2SO_4 through the process of biological oxidation by *Thiobacillus thiooxidens*. It is a very slow process. Due to this reason, the relative efficiency of mineral S is also not high.

$$2S + 5O_2 + 2\,H_2O \longrightarrow 2H_2SO_4 \quad (4.6)$$

Further reaction will follow as per eq. (4.3) and eq. (4.4) in the case of pyrites.

Sulphuric acid is applied directly to the soils. It constitutes one of the major chemical amendments in land reclamation programs in Russia. It reacts rapidly with soil lime since it does not have to go through the oxidation process. However, it is highly corrosive and quite difficult to handle at the field level.

Industrial Wastes

Phospho-gypsum

Phospho-gypsum (PG) is a byproduct of the phosphate fertilizer industry. The PG emanates when phosphoric acid is produced using rock phosphate through a process known as the wet process. The apatite (or rock phosphate) reacts with concentrated sulphuric acid to form phosphoric acid and phospho-gypsum as follows:

$$\underset{\text{Apatite (rock phosphate)}}{Ca_{10}(PO_4)_6F_2} + 10H_2SO_4 + 20H_2O \longrightarrow \underset{\text{Phospho-gypsum}}{10\,CaSO_4{\cdot}2H_2O} + 6H_3PO_4 + 2HF \qquad (4.7)$$

For each tonne of P_2O_5, 5.5 tonnes of byproduct PG, a fine powder like material containing about 25% moisture is produced. The pH is in the range of 4.5 and 5.0. PG contains 80-90% calcium sulphate and 1-2% fluorine. The PG sourced from M/s Paradeep Phosphate limited, Odisha has S 16.4%, Ca 18.8%, Na 0.11%, Al 0.02%, Fe 0.30%, Si 3.35% and P 0.34% (Nayak *et al.,* 2011). In comparison, mineral gypsum has S 16.1%, Ca 18.3%, Na 0.18%, Al 0.06%, Fe 0.01%, Si 0.82% and P 0.01%. The exact quality of PG depends upon the presence of impurities in the rock phosphate. Some of the impurities could be due to the presence of radioactive nuclide (Uranium and Radium) with minor amounts of toxic metals such as arsenic, barium, cadmium, chromium, lead, mercury, selenium, silver, fluoride and aluminum. Current researches revealed that one time application of phospho-gypsum for land reclamation and/or small doses for water reclamation are unlikely to result in phytotoxic level of fluoride. Similarly, fears of radioactive elements have also been discounted although the exact quantum and nature of these elements may vary due to the source of rock phosphate used in the manufacturing process. Studies so far have also proved that PG is a good amendment to reclaim alkali lands and alkali water.

Distillery spent wash

Distillery spent wash (DSW), a byproduct of the molasses based distilleries, is a nutrient rich wastewater. It has high biological oxygen and chemical oxygen demand (BOD and COD). It has 0.10-0.15% N, 0.01-0.03% P_2O_5, 0.80-1.20% K_2O and appreciable quantities of Ca, Mg,

S and micronutrients. It is estimated that for each liter of alcohol, about 15 liters of DSW is produced. In Tamil Nadu, under the AICRP on Management of Salt Affected Soils and Use of Saline Water in Agriculture, distillery spent wash (DSW) has been used to reclaim alkali lands. Around 5 lakhs liters of DSW needs to be applied to reclaim one ha area (Fig. 4.5; Haroon and Bose, 2004). DSW is also used as a source of nutrients, being rich in N, P and K.

Fig. 4.5 Application of DSW for reclamation of alkali lands

Press mud

Press mud is a byproduct of the sugar mills. Its output in India is increasing due to increasing numbers of sugar mills. It is estimated that currently 5.0 million tonnes of press mud is being produced annually. Sugar mills either use the carbonation (calcium carbonate-sulphur) or the sulphitation process (Sulphur) to purify the sugarcane juice. Press mud coming out of the sugar mills with the carbonation process contains about 70% calcium carbonate and 11% organic matter. Since the solubility of calcium carbonate is quite less, the press mud available through this process cannot be used as an effective amendment. However, it is applied along with farmyard manure as an amendment on mild alkali soils or to improve soil physical properties. The press mud, produced in sugar mills using sulphitation process, contains 9.3% calcium sulphate and 35.8% organic matter. Since the percent calcium sulphate is too less, the quantities required to reclaim barren lands would be too large. Nonetheless, it has a good potential to reclaim mild alkali lands as well as to ameliorate soils where sodicity build-up is due to irrigation with alkali water. Evidences have been generated to prove its usefulness as an amendment. Addition of 10-20 t/ha press mud can reduce the gypsum requirement of soils by about half because of the addition of calcium sulphate and organic matter.

Fly ash

Fly ash, a by-product of coal based thermal power plants, besides many elements contains soluble phosphorus, magnesium, sodium and calcium. Around 10 million tonnes of fly ash is generated by various thermal power plants in the country. Depending upon the origin of coal used in a particular power plant, fly ash could be alkali or acidic in nature. Few experiments have been conducted in India to establish its usefulness in alkali soils (Table 4.4). Guidelines for its application as an amendment are yet to be developed.

Table 4.4 Effect of fly ash and pyrites on rice and wheat yield in alkali soils

Amendment	Quantity (t/ha)	Yield (t/ha)			
		1979-80		1980-81	
		Rice	Wheat	Rice	Wheat
Control	0	1.02	0.57	1.98	0/91
Fly ash	10	2.85	2.40	4.03	0.75
Pyrite	10	3.03	2.87	4.33	2.79

Source: Tiwari *et al.* (1989)

Organic materials

Organic carbon content of most alkali soils is quite low compared to normal soils. Farmers have been applying organic manures such as farmyard manure, green manure and compost to these soils for a long time. Organic amendments like straw, rice husk (rice hulls), groundnut and safflower hulls, farmyard manure (FYM), compost, poultry droppings, green manure, tree leaves, saw dust etc. have also been used. Organic matter upon decomposition produces carbon dioxide and several organic acids. As such, soil pH is lowered and solubility of calcium carbonate and other soil minerals is enhanced leading to increase in the electrical conductivity of the soil solution. The enhanced water transport and increased calcium concentration of the soil solution helps to reduce the exchangeable sodium percentage of the soils. These reactions quite effectively occur under anaerobic conditions. Organic matter alone might not be effective in land reclamation. To be effective, relatively large quantities of organic amendments have to be used. Moreover, one has to keep in mind the C:N ratio, which varies widely. Saw dust, rice husk and barley straw decompose slowly and are less effective compared to *Sesbania* which has a narrow C:N ratio. In some cases, when C:N ratio is large, occurrence of N deficiency in soils cannot be discounted. The synergy of chemical and organic amendments has been found to help to reclaim the soils effectively and speedily. Besides, organic manures help to meet a part of the nutrient requirement of crops.

Relative Performance of Various Amendments

Application of amendment has profound positive effect on the physico-chemical properties of the soils and crops yield (Table 4.5). Since the cations replacement in soils is governed by the type of cation in the soil solution (its valency, size and degree of hydration), ion concentration, nature of ions associated with the cation and nature of colloidal particles, wide variations in soil properties might be realized. If we compare the relative change in some of the soil properties, it is noted that the final soil ESP with gypsum and H_2SO_4 as amendments is similar (Table 4.5). The H ion having the same valency as Na but being smaller in size and non-hydrated, it has greater replacing power and is preferred over Na on the exchange complex. Thus, H ions of the H_2SO_4 replace Na from the exchange complex to lower the soil ESP. In spite of ESP decrease being similar in H_2SO_4 and gypsum treatments, the saturated hydraulic conductivity of the soil was less and the water dispersible clay more in H_2SO_4 treated soil than gypsum treated soil (Table 4.5). Similarly, the ESP drop and improvement in soil physical properties was higher, when $Al_2(SO_4)_3$ was applied as an amendment. It could be due to Al of the $Al_2(SO_4)_3$, which is a more effective flocculent as compared to H of H_2SO_4. Similar comments are valid for Ca of $CaSO_4$. Besides solubility of $Al_2(SO_4)_3$ and $CaSO_4$ being low, major fraction of these amendments remain immobilized due to soluble $NaHCO_3$ and/or Na_2CO_3. As such, these amendments are quite effectively used to replace exchangeable sodium.

Table 4.5 Effect of different amendments (applied @100% GR) on physico-chemical properties of a sodic Vertisol

Amendment	pH (1:2)	EC_e (dS/ m)	ESP	Saturated hydraulic conductivity (mm/h)	Water dispersible clay (%)	Dispersion ratio
Control	8.81	0.80	65	0.06	37.2	0.41
Gypsum	7.9	0.72	14	4.77	8.0	0.09
Pyrites	8.0	0.31	20	1.64	32.4	0.36
H_2SO_4	7.5	0.18	14	2.98	30.4	0.54
$Al_2(SO_4)_3$	7.6	0.27	8	4.49	8.6	0.09
$FeSO_4$	7.9	0.85	21	1.59	33.7	0.37

Source: Sharma and Gupta, 1986

Pyrites versus gypsum

Since the relative performance of pyrites is related to initial water soluble S, the relative decrease in soil ESP and yield of rice and wheat crops with pyrites containing more than 5.5% water soluble sulphur gave almost equivalent results to that of gypsum applied at equivalent levels (Table 4.6, Table 4.7).

Gypsum versus phospho-gypsum

The studies at the Lucknow center of CSSRI revealed that phospho-gypsum is a better reclaiming agent as compared to mined gypsum. The PG treatment resulted in greater reduction in pH and ESP of both the surface and the lower layers. It is attributed to higher solubility of PG over mined gypsum, which allows soluble Ca to permeate to lower depths (Table 4.8). Application of phospho-gypsum also gave comparatively higher yield over equivalent dose of mined gypsum. Phospho-gypsum @ 10 t/ha (equivalent to 50% GR) resulted in slightly higher yield of 4.5 t/ha over the treatment 50% GRm (4.22 t/ha) and 25% GRm+25% GRpg (4.32 t/ha) in the first year of reclamation. Similarly, application of phospho-gypsum @ 5 t/ha resulted in comparatively higher yield over 12.5% GR_M +12.5% GRpg. The yield of follow up wheat were also higher in the PG treated fields (Fig. 4.6)

Table 4.6 Effect of gypsum and pyrites on yield and soil properties

Treatment	pH (1:2) (0-15 cm)		ESP (0-15 cm)	Yield (t/ha)	
	After paddy	After wheat		Paddy	Wheat
Control	10.0	10.1	69.1	3.3	0.0
Gypsum (50% GR)	9.5	9.3	50.2	5.2	3.3
Gypsum (75% GR)	9.3	9.3	32.9	5.3	4.2
Pyrite I* (50% GR)	9.9	9.9	60.2	4.6	0.6
Pyrites I (75% GR)	9.7	9.8	55.8	4.4	0.4
Pyrites II* (50% GR)	9.7	9.5	48.5	5.1	2.9
Pyrites II (75% GR)	9.2	9.2	30.9	5.3	4.1
Pyrites III* (50% GR)	9.7	9.7	50.0	5.6	3.4
Pyrites III (75% GR)	9.2	9.3	29.7	5.7	3.9

* Pyrites I, II and III contained 2.8, 5.5 and 8.0 soluble sulphur at the time of use

Source: Swarup and Yaduvanshi (2004)

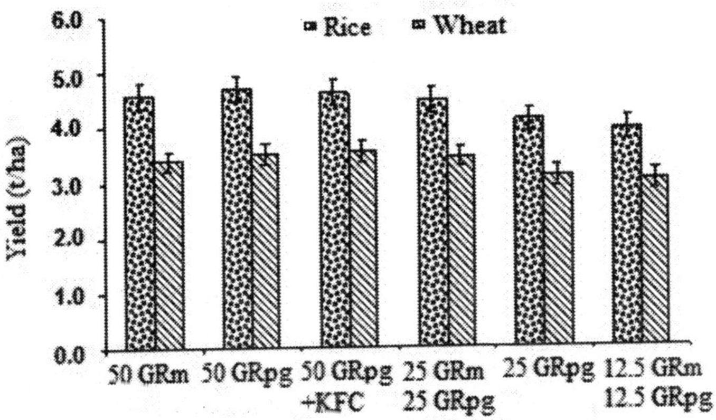

Fig. 4.6 Effect of gypsum and phospho-gypsum on rice and wheat yield at Shivri, Lucknow

Table 4.7 Effect of gypsum and pyrites on the yield of crops in alkali soils

Source	Soil type and pH	Crops	Grain yield (t/ha) at different levels (% GR) of amendments									
			0		25		50		75		100	
			G	P	G	P	G	P	G	P	G	P
Abrol and Verma (1978)	Alfisol, initial pH 10.5	Rice	3.86	3.86	6.71	5.71	6.81	6.09	7.44	6.71	7.24	6.91
		Wheat	0.02	0.02	0.15	0.15	3.15	0.54	3.60	1.35	4.22	1.35
Chauhan (1978)	Alfisol, initial pH 9.2	Rice	1.73	1.73	-	-	2.77	0.54	-	-	-	-
		Barley	2.62	2.62	-	-	3.67	3.64	-	-	-	-
Bhatt *et al.* (1978)	Alfisol, farmers field, pH varying from 9.0 to 10.5	Rice	2.93	2.93	3.23	3.49	3.57	3.87	3.13	4.06	4.09	4.87
		Barley	1.50	1.50	1.18	1.18	1.74	1.27	1.58	1.60	2.33	1.86
Singh *et al.* (1981)	Alfisol, initial pH 10.2	Rice	2.61	2.61	-	6.65[1]	6.87	-	-	7.59[2]	-	7.66
		Wheat	2.05	2.05	-	3.72	4.80	-	-	4.84	-	5.00
Sharma and Gupta (1986)	Vertisol, initial pH 8.5	Rice	0.14	0.14	0.97	0.84	1.77	1.60	2.50	2.15	-	-
		Wheat	0.50	0.50	1.00	0.88	1.67	1.51	2.40	3.22	-	-
Chhabra *et al.* (1989)	Alfisol, initial pH 10.2	Rice	1.75	1.75	2.93	2.09	3.06	2.37	-	-	3.43	2.60

G= Gypsum: P= Pyrites: 1- 33% GR, 2- 66% GR.

Table 4.8 Effect of phospho-gypsum on ESP of soil after first and third year of rice-wheat cropping

Treatments	2006-07		2008-09	
	0-15 cm	15-30 cm	0-15 cm	15-30 cm
50GRm	26.5	31.6	24.6	28.5
50GRpg	20.4	27.3	18.3	25.7
50GR$_{PG}$-KFC	18.3	25.2	17.9	23.4
25GRm+ 25GRpg	17.8	23.5	16.7	21.5
25 GRpg	19.4	25.6	17.9	24.3
12.5GRm+ 12.5GRpg	25.2	30.8	23.5	27.5
LSD (p = 0.05)	1.5	2	1.5	2

50GRm = 50% gypsum requirement through mineral gypsum;
50 GRpg = 50% gypsum requirement through phospho-gypsum;
25 and 12.5 means 25% and 12.5% of GR

Organic Amendments

Singh (1974) observed that when barley straw and *Sesbania* were decomposed under aerobic conditions, the reduction in ESP and improvements in other soil physical properties were less compared to when decomposed under saturated (anaerobic) conditions (Table 4.9). Organic materials together with inorganic amendments are more cost effective, hasten the reclamation process, increase the yield and therefore, their use should be encouraged (Dargan *et al.*, 1976).

Table 4.9 Effect of C: N ratio of added organic matter @ 2% and soil saturation on improvements in physico-chemical properties of a calcareous alkali silty clay soil

Treatment	pH_s	ESP	Modulus of rupture (bars)	Hydraulic conductivity (cm/h)	Na in leachate (meq/L)
Barely straw aerated	9.1	81.5	12.9	0.03	16.6.
Sesbania straw aerated	8.8	68.5	11.8	0.04	26.3
Barley straw saturated	8.4	56.1	11.3	1.72	30.2
Sesbania straw saturated	8.3	47.2	7.5	1.72	30.2
Original soils	9.5	95.0	-	-	-

Source: Singh, 1974

QUANTITY OF AMENDMENT

The amount of an amendment for reclamation depends upon the amount of exchangeable sodium to be replaced, which in turn is governed by the amount of sodium adsorbed on to the soil (ESP and CEC), sodicity tolerance and rooting depth of the crop.

Gypsum Requirement

Gypsum requirement (GR) for reclamation of alkali soils is determined in the laboratory or by several other means. Theoretically, for every one milliequivalent of Na to be replaced one needs one me of soluble Ca. This will amount to 0.02 g Ca /100 g soil or 0.86 g of pure $CaSO_4.2H_2O$/kg. Taking 2 x 10^6 kg as weight of 0-15 cm soil layer, the amount will be 1.72 tonnes of gypsum per ha per me of Na to be replaced. Thus, the GR can be calculated by the following relationship:

$$GR\ (meq/100\ g) = [CEC\ (initial\ ESP - final\ ESP)]\ /\ 100 \qquad (4.7)$$

$$\text{GR/ ha-15 cm} = \text{GR (meq /100 g) } 860 \times 10^{-6} \times 2.10^{6}$$
$$= \text{GR (meq /100 g)} \times 1720 \text{ (kg)} \quad (4.8)$$

For reclaiming upper 30 cm soil one may need about double the amount of above calculated gypsum if the initial ESP of this layer is of similar magnitude as that of 0-15 cm layer. If the gypsum is less than 100% pure, the quantity of gypsum needs to be increased proportionately.

In the laboratory, Schoonover method is used to work out the gypsum requirement (Schoonover, 1952). As per this procedure, one needs to prepare a saturated gypsum solution of known calcium concentration by adding 5 g of chemically pure $CaSO_4$-$2H_2O$ to one liter of distilled water. Shake vigorously for about 10 min on a mechanical shaker or 1 h by hand. Filter through Whatman filter paper no. 1.

Take 5 g of air dry soil in 250 ml conical flask. Add 100 ml of the saturated gypsum solution. Firmly put a stopper and shake for 5 min on a mechanical shaker or 50 min by hand. Filter the contents through Whatman filter paper no. 1. Transfer 5 ml of aliquot into 100 ml porcelain dish. Determine the calcium and magnesium concentration. Run a blank using 5 ml of the saturated gypsum solution prepared before.

GR (meq/100 g soil) = 2 [(Ca concentration in blank, meq/L) – (Ca concentration in the filtrate, meq/L)]

GR for 15 cm soil layer (t/ha) = GR (meq/100 g soil) * 1.72 (4.9)

Gypsum requirement (GR) of a barren soil determined by this method is quite high. Abrol *et al.* (1975) argued that the gypsum requirement as calculated by Schoonover method, also includes the gypsum required to neutralize the soluble sodium carbonate, if present in alkali soils. It is therefore, prudent to leach the sample of the soil (5 g) with 20 ml of 60% ethanol. It will leach down the soluble sodium carbonate and therefore, will not add to the gypsum requirement of the soil. Rest of the procedure is same as discussed in the foregoing paragraph. Moreover, the GR calculated for the top 30 cm of the soil could be reduced by half considering that only top 15 cm needs to be reclaimed initially. Extensive work done at Central Soil Salinity Research Institute, Karnal, India, has shown that 10 to 15 t/ha of gypsum containing 70% $CaSO_4.2H_2O$, which approximates about half of the actual GR determined by Schoonover's method, is sufficient to reclaim the surface 15 cm soil if one is to start with rice as the first crop. This is keeping in view the high tolerance of transplanted rice to soil ESP and its shallow root system. Moreover, its roots capacity to facilitate the soil water to dissolve soil $CaCO_3$ also helps to reduce the soil exchangeable sodium. Over time, ESP of the surface as well as deeper layers gets reduced so as to allow cultivation of other crops. To grow sensitive crops like pulses in the initial years, one needs to add high amounts of amendments in the initial years of reclamation.

pH versus GR Graph

The absolute quantity of exchangeable sodium percentage of an alkali soil will continue to be the basis to accurately determine the quantity of gypsum required for reclamation. In practice, pH of the saturated soil paste (pH_s) or pH of 1:2 soil water suspension (pH_2) may also be used since both are fairly correlated with each other and with ESP of the alluvial soils of the Indo-Gangetic plains. This method is particularly useful where skilled staff and laboratory facilities to determine the exchangeable sodium and gypsum requirement are not available. Abrol *et al.* (1973) prepared a graph between pH_2 and GR of the soils for different soil textures (Fig. 4.7). This graph is a quick and fairly good method to estimate GR of alkali soils of the Indo-Gangetic plains. The GR calculated with this graph should not be reduced to half as for laboratory determined GR. It may however, be noted that this graph cannot be used universally for all kinds of soils. It is observed that pH_s and ESP of black cotton soils (Vertisols) are not well related and therefore, use of such a graph will not be advisable for such soils. Once the GR is known, equivalent quantities of other amendments could be easily calculated using Table 4.9.

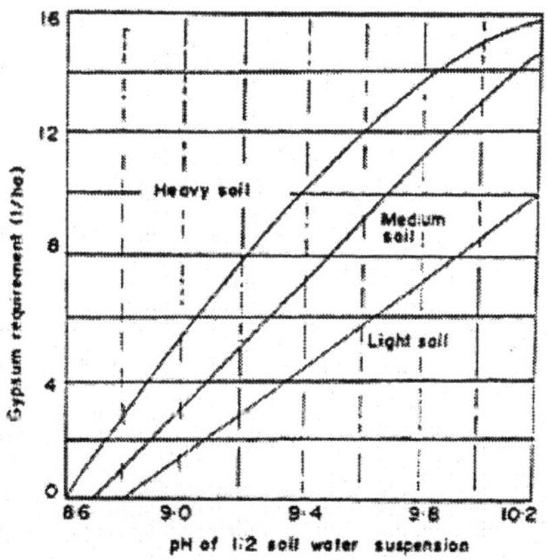

Fig. 4.7 pH_2 versus GR graph for alluvial soils on Indo-Gangetic plains

Table 4.9 Commonly used amendments and their equivalent amendment values

Amendments (Chemical formula)	Tonnes of amendment material equivalent to	
	1 tonne of pure gypsum	1 tonne of pure sulphur
Gypsum ($CaSO_4 . 2 H_2O$)	1.00	5.38
Calcium chloride ($CaCl_2 . 2 H_2O$)	0.86	--
Calcium nitrate ($Ca (NO_3) . 2 H_2O$)	1.06	--
Press mud (lime–sulphur, 9% Ca+24% S)	0.78	4.17
Sulphuric acid (H_2SO_4)	0.61	3.20
Iron sulphate ($FeSO_4 . 7 H_2O$)	1.62	
Ferric sulphate ($Fe_2SO_4)_3$.	1.09	5.85
Aluminium sulphate ($Al_2 (SO_4)_3 . 18 H_2O$)	1.29	6.34
Sulphur (S) **	0.19	1.00
Pyrites ((FeS_2 . 30% S) **	0.63	

*The above quantities are based on 100% pure materials. If the purity is less than 100%, then necessary corrections must be made. Thus, if gypsum is only 80% pure then the equivalent quantity to be applied would be 1.25 times of quantities mentioned in the Table.

** For sulphur and pyrites, 100% oxidation is assumed.

The application of graded doses of gypsum in rice-wheat cropping revealed that only half of the laboratory determined GR needs to be applied to get good yields (Table 4.10). For high efficiency of applied gypsum, following points should be kept in view:

- All alkali soils contain measurable amount of $NaHCO_3$ and Na_2CO_3 in the soil solution, which under normal conditions may react with the gypsum to neutralize and precipitate soluble Ca before it is used to replace exchangeable Na. To avoid this reaction, it is recommended that a heavy pre-irrigation should be applied before gypsum application
- Spread the gypsum uniformly over the whole area
- Fineness of gypsum should conform to the Bureau of Indian Standards (BIS) specifications i.e. the whole of the material must pass through 10 mesh and at least 50% of it must pass through 60 mesh (0.25 mm) sieve. This standard pertains to 30

mesh fineness. The gypsum supplied by State Land Reclamation Boards generally meets this standard

- The gypsum should be mixed within the shallow depth of about 10 cm (Table 4.11). With deeper mixing, one ends up with lesser crop yields at least in the first two years.

Table 4.10 Yield (t/ha) of crops as affected by gypsum treatment of an alkali soil

Gypsum (t/ ha)	*Sesbania** 1970	Wheat 1970-71	Rice 1971	Wheat 1971-72	Rice 1972	Wheat 1972-73	Mash 1973	Lentil 1973-74	Pearl millet 1974	Gram 1974-75
0.0	0.44	0.0	4.39	1.52	7.18	1.35	0.05	0.28	1.06	0.12
7.5	3.99	1.89	6.21	2.79	7.23	1.96	0.25	0.82	1.33	0.47
15.0	4.35	3.49	6.39	3.45	7.13	2.43	0.32	1.16	1.96	0.69
22.5	4.70	4.16	7.08	3.72	7.17	2.26	0.38	1.36	1.81	0.90
30.0	4.50	3.79	6.66	3.98	7.03	2.74	0.42	1.40	1.94	0.93
LSD at P = 0.05	0.51	0.64	0.81	0.70	NS	0.50	0.12	0.35	0.52	0.44
Normal yield	-	4.5	7.0	4.5	7.0	4.5	1.0	2.0	3.0	2.0

* Green matter yield; Initial soil pH = 10.3; ESP = 94 in surface 15 cm soil
Source: Abrol and Bhumbla (1979)

Table 4.11 Grain yield of crops as affected by quantity of gypsum and depth of mixing in an alkali soil

Gypsum level (t/ha)	Mixing depth (cm)	Grain yield (t/ha)		
		Barley	Rice	Wheat
13.5	10	2.64	7.02	3.28
	20	2.46	6.05	3.16
	30	0.53	5.64	2.00
27	10	2.83	6.71	3.67
	20	3.22	6.28	3.67
	30	2.30	5.98	3.58

Source: Khosla *et al.* (1973)

LEACHING

To transport the chemical constituents of an amendment to the exchange sites and to leach down the exchanged chemical constituents from the root zone, it is imperative to carryout leaching. It helps to leach down the soluble salts and/or the exchanged products below the root zone by passage of water through the soil. It is accomplished by ponding water on the soil surface so that water percolating through the root zone dissolves the soluble salts and exchanged ions and takes them along with it below the root zone.

Since the solubility of gypsum in pure water is quite less, it was thought that water required to leach gypsum amended soils would be quite high. On the contrary, it is now known that depth of water required to dissolve gypsum in soils of high ESP would be much less than calculated on the basis of its solubility in pure water. It is due to the fact that soil acts as a sink for the calcium ions. Normally, 50 to 80% of the applied gypsum gets dissolved within 7 to 10 days of continuous ponding and leaching. There is no need to leach down the soils beyond this period as the remaining leaching requirement is met by the ponded water in the rice crop. Since rice is the first crop in land reclamation programs, this recommendation is valid. If crops other than rice are cultivated, leaching period could be increased.

In pyrites or S treated soils wetting and drying cycles should help to maintain proper moisture and aerobic conditions necessary to ensure maximum oxidation and production of H_2SO_4. For pyrites amended soils, fields are left as such for about 7-10 days after applying the amendment. Afterwards, leaching is carried out for about 7-15 days either in continuous ponding mode or in intermittent ponding mode. Water standing in rice fields would take acre of the remaining water requirement for leaching in this case as well.

Flushing

Leaching process is greatly inhibited in heavy soils or in alkali soils with low infiltration rates and having shallow water table. In such events, flushing is adopted as a means of leaching. Flushing of the ponded water helps to remove the surface concentrated salts. The studies conducted in an alkali soil indicated that irrigating the plowed soil, turning of soil with a power tiller and flushing of standing water after 24 h followed by a second flushing after 7 days helped to leach down the salts before gypsum application and leaching (Fig. 4.8). Compared with conventional treatment, the surface flushing significantly reduced the EC_e and SAR of the soil and improved physical properties such as f-potential, dispersible clay content, water stable aggregates and saturated hydraulic conductivity. With flushing, average yields of rice and wheat increased by 25% and 62%, respectively as compared to the conventional practice of leaching alone. Split application of gypsum between flushing was not beneficial and therefore, full dose of gypsum should be applied after second flush.

Fig. 4.8 Fields being filled with water for flushing

SELECTION OF CROPS AND CROPPING SEQUENCES

Crops differ widely in tolerance to alkali environment of the root zone. Even crop varieties respond differently to alkali conditions. Therefore, selection of crops and cropping sequences is of practical and great economic importance to manage alkali lands. Although the crop tolerance to soil sodicity might vary with the stage of growth, variety, environmental factors and level of management, their relative tolerance is summarized in Table 4.12 and 4.13.

Table 4.12 Tolerance of crops to various levels of ESP

Range of ESP	Crops
10 – 15	Safflower, mash, peas, lentil, pigeon pea, *urd* bean
15 –20	Bengal gram, soybean, maize
20 – 25	Groundnut, cow pea, onion, pearl millet
25 – 30	Linseed, garlic, guar, lemon grass, pamarosa, sugarcane, cotton
30 – 50	Wheat, *raya*, sorghum, sunflower, *berseem*, shaftal, *senji*, blue panic, gutton panic
50 – 60	Barley, *sesbania*, para grass, Rhodes grass, matricaria
60 - 70	Rice, sugar beet, Karnal grass

Note: Relative crop yields are only 50% of the maximum in the indicated ESP range

Table 4.13 Relative tolerance of crops and grasses to exchangeable sodium

Tolerant	Moderately tolerant	Sensitive
Karnal grass (*Leptochloa fusca*)	Wheat (*Triticum astivum*)	Gram (*Cicar arielimun*)
Rhodes grass (*Chloris gauana*)	Barley (*Hordecum valgaro*)	*Mash* (*Phaseolus mungo*)
Para grass (*Brachiaria mutica*)	Oat (*Avena sativa*)	Chickpea (*Cajanus cajan*)
Bermuda grass (*Cynodon dactylon*)	*Shaftal* (*Torifolium resupinatum*)	Lentil (*Lens esculenta*)
Rice (*Oryza sativa*)	Lucern (*Medicago sativa*)	Soybean (*Giycino max*)
Dhaincha (*Sesbania aculeate*)	Turnip (*Brassica rapa*)	Groundnut (*Archis hypogaea*)
Sugar beet (*Beta vulgaris*)	Sunflower (*Helianthus annus*)	Sesamum (*Sesamum oriental*)
Teosinte (*Euchlaona maxicana*)	Safflower (*Carthamus tinctorius*)	*Moong* (*Phaseolus aureus*)
	Linseed (*Linum usitaissum*)	Pea (*Pisum saccharatum*)
	Onion (*Allium cepa*)	Cowpea (*Vigna unguiculata*)
	Garlic (*Allium sativum*)	Maize (*Zea mays*)
	Pearl millet (*Benisetum typhoites*)	Cotton (*Gosstypium fndicum*)
	Cotton (*Gosssypium indicum*)	

Choice of the tolerant varieties within a crop can result in more production and better economic returns from alkali soils (Mishra, 1996). Such varieties have been identified and evolved through selection and hybridization in crops like rice, wheat, mustard and gram (Table 4.14). It may be noted that some variations might occur in different tables reported by different workers. It is mainly because of the variations in the experimental conditions, varieties used and other agro-climatic conditions. For example, transplanted lowland rice is more tolerant to alkali stress than direct seeded rice many times causing confusion on its tolerance to alkali conditions. Crops that can withstand excess moisture conditions are generally more tolerant to alkali conditions.

Cropping Pattern

In order to earn more and to initiate a process that helps to reclaim the soil, an appropriate cropping sequence must be selected. Results of various experiments on comparison of different cropping systems such as rice-wheat, rice-*berseem,* rice-mustard, sorghum-wheat, sorghum-mustard and sorghum-berseem at Karnal indicated that the rice based cropping patterns are more promising and profitable than sorghum based sequences (Table 4.15).

Complete reclamation of alkali soils takes several years. After following rice-wheat or rice-*berseem* sequence for about 3 years, diversification of cropping pattern can be considered depending upon the extent of land reclamation (Table 4.16). With passage of time more sensitive crops could be chosen such that after 8-10 years any crop could be grown on the reclaimed lands.

Table 4.14 Promising varieties of important crops for cultivation in alkali soils

Crop	Tolerant varieties	Soil pH_2
Rice	CSR 10*, CSR 11, CSR 12, CSR 13*	9.8 -10.2
	CSR19, CSR23*, CSR27*, Basmati CSR30*, CSR36*	9.4-9.8
Wheat	KRL 1-4*, WH157, KRL 210*, KRL 213*	<9.3
	Raj3077, KRL19*	<9.3
Barley	DL200, Ratna, BH97, DL348	8.8-9.3
Indian mustard (Raya)	Pusa Bold, Varuna	8.8-9.2
	Kranti, CS52*, CSTR330-1	8.8-9.3
	CST609-B 10, CS54*, CS56*	8.8-9.3
Gram	Karnal Chana 1	<9.0
Sugarbeet	Ramonskaaya 06, Maribo Resistapoly	9.5-10
Sugarcane	Co453, Co1341	<9.0

*Institute varieties released by Central Varietal Release Committee

Table 4.15 Cropping patterns recommended for alkali soils (reclamation phase)

Season	Year of reclamation		
	First	**Second**	**Third**
Summer (April – May)	Leaching/ *Sesbania* green manuring	*Sesbania* green manuring	*Sesbania* green manuring/ Fallow/Cow pea (fodder)/ Maize (fodder)
Kharif	Rice/ *Sesbania*	Rice	Rice
Rabi	Wheat/Barley	Wheat/ Barley/ Raya/Shaftal Senji/ Sugar beet	Wheat/Barley/Raya/Berseem/ Senji/Sugar beet

NUTRIENT MANAGEMENT

High pH, high exchangeable sodium, high $CaCO_3$, negligible to low organic matter content and inhospitable soil physical environment of alkali soils affect the transformation and availability of native and applied fertilizer nutrients (Gupta and Abrol, 1990). Alkali soils, in general, are deficient in Ca, N and Zn, but are rich in available P and K. Following practices are recommended for efficient nutrients management in alkali soils:

Table 4.16 Cropping sequences recommended for alkali soils (Post reclamation phase)

Soil pH_2 (0-15 cm layer)	Years after reclamation	Crops	
		Kharif	*Rabi*
9.2 to 9.3	3 to 4	Cotton, sorghum, pearl millet, rape seed	Mustard
8.9 to 9.0	5 to 8	Sugarcane, groundnut, pigeon pea, soybean	*Berseem*, sunflower, safflower, gram, pea, lentil
8.5 to 8.8	After 8 to 10 years	Any kind of crops including vegetables, flowers etc.	

- Practise green manuring with *sesbania* every year or at least once in two years to improve organic matter status and to overcome nitrogen deficiency problem.
- Apply 20 to 25% more nitrogen to rice-wheat in the initial years than recommended for normal soils. Apply 150 kg N/ha through urea or ammonium sulphate in 3 to 4 splits to compensate for high volatilization losses.
- Since alkali soils are generally rich in P and K; their application could be withheld for three to four years. After this, apply P only to rice crop preferably on soil test basis. Singh (1998) reported that rice crop responds to phosphorus when the Olsen's P of the surface soil falls below 7.5 kg/ha. The wheat yields do not show decline even at this critical level for the next few years. Application of P to either one or both crops significantly increased the grain yield of rice only after 5 years of cropping and of wheat after 11 years.
- Application of Zn plays an ameliorative role in alkali soils by selective absorption enhancing Ca^{++}/Na^+ and K^+/Na^+ ratios in plants. Application of 10 kg $ZnSO_4$/ha to each crop of rice and wheat or 20 kg $ZnSO_4$ /ha to rice increased the grain yield significantly over no Zn (Singh and Abrol, 1986). Thus, apply 10 to 20 kg $ZnSO_4$/ha for initial two to three years both to rice and wheat crops. Later on, depending upon soil test values, zinc may be applied to rice only.

- Symptoms of manganese deficiency have recently been reported in wheat crop grown on reclaimed calcareous and light textured alkali soils. To overcome the deficiency, $MnSO_4$ @ 10 kg/ha should be applied through soil or 0.5% solution may be sprayed when such symptoms are noticed.

Integrated Nutrient Management

Disproportionate and inappropriate continuous application of chemical fertilizers has lowered the fertility status and caused nutrient hunger in soils. Crops yields have either stabilized or even declined over the years. Integrated nutrient management comprising of a good blend of organic and chemical fertilizers helps to sustain crop production in reclaimed lands. Besides, farmyard and green manure increase organic carbon and also increase the availability of plant nutrients. FYM also helps to meet a part of the zinc requirement of the crops. Evidences exist to show that application of farmyard manure and/or green manure helps to build-up nitrogen status of the soils and over the years it increases the yields of crops. Experiments in gypsum amended alkali soils revealed that about 110 kg nitrogen and 11 kg of phosphorus is made available when 50 days old *dhaincha* crop is turned into the soil. As a result, rice and wheat yield increased by 1.48 and 0.67 t/ha respectively.

Yield of rice and wheat significantly increased with integrated use of green manuring or FYM at 100% recommended dose of chemical fertilizers (120 N, 26 kg P and 42 kg K/ha) as compared to 100% recommended inorganic fertilizers alone (Table 4.17). The yields in integrated nutrient management treatments were nearly equal to the yield obtained with N180 P39 K36 applied solely through chemical fertilizers.

Table 4.17 Grain yield of rice-wheat on sodic soil with or without integrated use of fertilizers

Treatment	Grain yield (t/ha)									
	1994-95		1995-96		1996-97		1997-98		1998-1999	
	Rice	Wheat	Rice	Wheat	Rice	Wheat	Rice	Wheat	Rice	Wheat
N_{120} P_{26} K_{42}	5.57	5.00	5.50	5.16	5.52	5.05	5.57	4.96	5.45	4.87
N_{120} P_{26} K_{42} + GM	6.25	5.30	6.35	5.28	6.98	5.29	6.94	5.57	6.65	5.41
N_{120} P_{26} K_{42} + FYM	6.12	5.15	6.15	5.10	6.56	5.14	6.67	5.25	6.59	5.52
CD (P=0.05)	0.52	0.52	0.58	0.51	0.56	0.40	0.56	0.50	0.33	0.53

IMPROVED CULTURAL PRACTICES

Plant Population

Germination is generally inhibited in alkali soils due to formation of a hard crust or other adverse physico-chemical properties. High rate of mortality especially during early stages further depresses the plant population. This together with poor tillering reduces the yield of the crops. Increasing the plant population either by reducing the plant-to-plant and row-to-row distance in transplanted crops or increasing the seed rate in seeded crops helps to improve crop stand in alkali soils (Chhabra, 1996). In rice, increasing the number of seedlings per hill and by gap filling also helps to achieve higher yields.

Puddling

Puddling in lowland rice cultivation is carried out mainly to reduce the infiltration rate to save water and to control weeds. In alkali soils, infiltration rate is low and weed problem is not encountered at least during the first 2-3 years. Therefore, puddling can be avoided in these soils unless the soil is too hard to transplant the rice seedlings. Puddling in that situation can be done to facilitate transplanting.

Age of Seedlings

The tolerance of crops, in general, increases with age. In rice, older seedlings (40-45 days) establish better than younger seedlings (25-30 days). Moreover, it is better to grow the rice seedling in normal soils as it would provide healthy seedlings for transplanting. To establish tree species, planting older seedlings grown in normal soil has proved beneficial.

Green Manuring

Application of green manure enhances OM content of the soil, increases partial pressure of CO_2, lowers pH, enhances solubility of native $CaCO_3$ and adds considerable amount of plant nutrients to the soil. *Sesbania*, a GM crop, tolerant to high ESP and water logging, is an ideal crop. It is grown during summer months (May-June), a lean period in the rice-wheat cropping sequence. As such, it fits well into this cropping pattern.

Normally, 45-50 days old *Sesbania* crop, when it has attained a height of 1.5 to 1.8 m, is incorporated into the soil. Longer growth period renders the plant hardy and its decomposition becomes difficult. Having a narrow C:N ratio (about 25), it decomposes very easily and quickly. Rice seedlings can be transplanted soon after the incorporation as no decomposition period is needed. The standing water in rice fields hastens the decay without any harm to rice crop. Allowing more time for decomposition of a succulent crop like *Sesbania* results in greater loss of N through volatilization and N recovery by the crop is less. Green manuring ensures steady supply of N to the rice crop, since about 50% of the organic nitrogen contained in the green manured crop is converted into the readily available ammonical form within 4 to 6 days and the rest in another 10 to 20 days of incorporation. The yield increase in rice crop due to green manuring is equivalent to about 80 kg of N applied through chemical fertilizers. Moreover, it saves about 10-15 days as rice crop is transplanted straightway after incorporation of the green manure.

Water Management

Irrigation management in alkali soil poses peculiar problems due to clogging action of dispersed soil particles and low stability of the soil aggregates, which affects the water and air permeability. Because of the restricted entry of water, depleted soil moisture is not recharged fully during irrigation. Moreover, the amount of available water held by the soil between field capacity (0.1 bar) and wilting point (15 bars) goes on reducing with increasing ESP (Acharya and Abrol, 1991). Thus, the total water storage capacity of the soil decreases with increase in exchangeable sodium. As the depth of irrigation water and interval between two irrigations depends upon the evaporative demand, nature of plant response to soil water stress, depth of root development and the water storage and transmission characteristics of the soil, light and frequent irrigations should be applied in alkali soils. The total amount of irrigation water remaining the same; the water should be applied frequently in small quantities. Heavy irrigations can cause temporary water logging and oxygen stress in the root zone, which would affect the crop growth/yield adversely.

Moreover, in early stages of reclamation, bulk of the roots is confined to only surface 30 cm layer. Thus, in addition to the limitations imposed by physical properties of the soil, limited root growth restricts the moisture extraction from deeper layers following irrigation. All these constraints adversely impact the ability of a reclaimed alkali soil to supply adequate amount of water to the growing plants.

During the post-irrigation period also, the water transmission from the lower to the upper layers is severely restricted due to low hydraulic conductivity of the soil. Acharya *et al.* (1979) observed that with increasing ESP of the surface soil, the water uptake by *Brassica juncea* reduced drastically (Table 4.18). Not only that, the contribution of lower layers was low and practically none when the ESP was high. Thus, even if the water content is same, the available water is much less in an alkali than a normal soil (Fig. 4.9).

Table 4.18 Water extraction by *Brassica juncea* from different depths as related to the ESP of the surface soils

Water extraction	ESP of surface 15 cm soil				
	4	**11**	**16**	**23**	**18**
Extraction rate (cm/d)	0.23	0.24	0.20	0.18	0.14
Soil layer depth (cm)		Per cent of total extraction			
0-15	46.4	40.0	67.5	75.0	78.0
15-30	30.0	31.1	28.0	25.0	22.4
30-45	16.3	18.7	4.5	0	0
45-60	7.3	6.2	0	0	0

*Mean of 17 days extraction period; Source: Acharya *et al.* (1979)

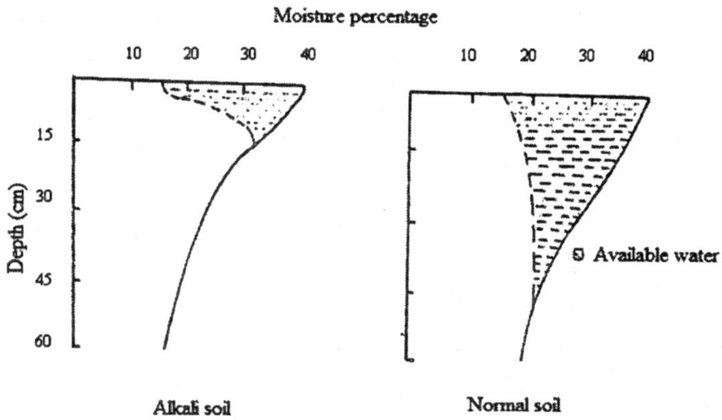

Fig. 4.9 The variation of available water in an alkali and a normal soil

Sharma *et al.* (1985) reported that compared to 21 days in a normal soil, the first irrigation to wheat should be given at 30 days after sowing in a alkali soil (ESP = 28 to 50). The maximum yield of wheat was recorded when 5 irrigations were given at different crop growth stages in rice-wheat cropping system (Table 4.19). Maximum water use efficiency of 6.5 kg grain/mm of water was recorded when irrigation was given at 1.0 IW/CPE ratio than the other treatments (Sharma *et al.*, 1990). Sharma and Singh (1992) reported that maximum yield of mustard under rice-mustard cropping system in alkali soils is obtained when water stress is avoided at rosette and siliquae formation stages.

Table 4.19. Yield and water use efficiency of wheat as influenced by different treatments in an alkali soil

Treatments	Grain yield (t/ha)	% yield compared to yield with 5 irrigations	Water use efficiency (kg/ha-mm)
Rainfed	4.4	14.8	2.65
One (CRI)	11.8	39.7	5.38
Two (CRI+F)	17.2	58.0	6.24
Three (CRI+J+M)	21.6	72.8	6.57
Three (CRI+T+M)	25.0	84.6	7.53
Three (CRI+T+F)	21.9	74.0	6.70
Four (CRI+T+J+M)	27.7	93.5	7.23
Four (CRI+J+F+M)	26.6	88.5	6.96
Five (CRI+T+J+M+D)	29.6	100	6.78
CD at 5 %	1.53	-	0.12

CRI = crown root initiation, T = Tillering, J = Jointing, F = Flowering, B = Booting, M = Milk formation and D = Dough stage

From practical standpoint, Dhruva Narayana (1980) reported that in tube well irrigated areas where tube well discharge is around 10 L/s, the border for rice-wheat cultivation could be 60-70 m in length and 6-8 m wide. Similarly, Tyagi (1976) reported that the graded border with a grade of 0.1-0.2% perform better than level borders. The highest application, distribution and water use efficiencies in a graded border could be achieved with a cut-off ratio of 70% except for the first irrigation where cut off ratio should be 85%.

RESODIFICATION OF RECLAIMED ALKALI LANDS

With the initiation of land reclamation process and because of the roots' activities especially rice, the pH of the soil both at the macro and micro levels decreases sharply. It increases the solubility of the native

$CaCO_3$ and decreases the dispersion of the organic matter (Chhabra and Abrol, 1977). These processes help in self reclamation of the soils both at top as well as down through the profile. As such, gypsum is applied only once during the first year. No further addition of gypsum is needed once the land is reclaimed. On the contrary, many farmers have reported resodification of the reclaimed soils. In Uttar Pradesh, large areas have been reported to have lost their productivity because of the resodification. Extensive surveys have revealed that such a situation has arisen mainly due to mismanagement of the reclaimed lands or due to the addition of Na from outside sources. On the basis of these surveys, following reasons could be cited for the resodification:

- Entry of runoff from unreclaimed alkali lands
- Use of high RSC water for irrigation
- Shallow water table more so of high residual sodium carbonate
- Fallowing of the reclaimed land for prolonged periods.

It has emerged that in Haryana, use of high RSC water and in Uttar Pradesh shallow water table and leaving the land fallow are major factors causing resodification.

OTHER MODELS FOR LAND RECLAMATION

Besides the main chemical amendment based model, several other models have been developed keeping in view the site specific requirements and socio-economic constraints of the clients. Some of these models are briefly discussed in the following sections.

Physical Reclamation

Profile studies have revealed that soil ESP decreases with depth at many places. Areas in Sharda *Sahayak* canal command especially situated on both sides of the main canal are few such locations. At such places, it should be possible to partially reclaim the soil through soil inversion. Depending upon the degree of variation, soil inversion with mould board or disc plow could bring low ESP soils at the top. It could reduce the gypsum requirement of the soil to a greater extent. In Uttar Pradesh, this model was tested while implementing a multi-enterprise model. The soil inversion played a significant role in land reclamation. While digging the aquaculture pond, the upper high ESP soil went down while the low ESP soil came at the top (Fig. 4.10). Thus, crops on the dykes and raised land could be cultivated without any chemical amendment.

Taking advantage of the above situation, a land reclamation model based on the concept of land modifications (physical land reclamation) and pond based integrated farming system was conceptualized and tested at farmers' field near Sharda Sahayak Canal at Kashrawan village in Raibareli district of UP. The results showed that (i) waterlogged alkali soil along the Sharda Sahayak canal command can be reclaimed and put to economic use by pond based farming system without application of gypsum (ii) on an average, one-meter seepage water level in the pond could be maintained for aquaculture without addition of water from external sources. Thus, aquaculture activity can be promoted along the canal with intercepted seepage which otherwise results in the rise in ground water table, (iii) the quality of the seepage water harvested in the pond was suitable for fish production, (iv) besides rice- wheat cropping system, vegetables and fruits could be raised in the very first year under this system, which was not otherwise possible with conventional method of land reclamation and (v) the productivity of the waste land and the productive use of canal seepage water are likely to increase over the years. A secondary benefit of this system would be realized over the years as the lowered water table would help to reclaim additional areas. This model quite suits the land holding pattern of the small and marginal farmers and has the potential to restore and support the livelihood through enhanced regular income and year round employment. The income from various enterprises and for the overall model depicted in Table 4.20 reveal highest B:C ratio for aquaculture followed by vegetables (Sharma *et al.*, 2008, Mishra *et al.*, 2012).

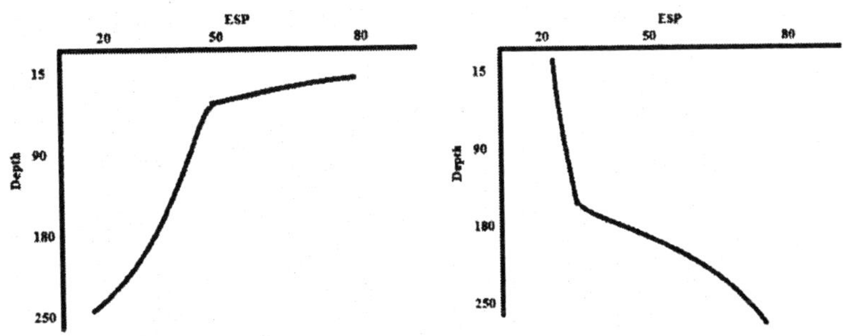

Fig. 4.10 The schematic ESP of the soil before inversion (left) and after inversion (right)

Biological Reclamation Model for Resource Poor Farmers

The biological reclamation approach aims at the reclamation of alkali soils by growing salt tolerant crops and their varieties. It either exploits salt tolerant grasses as the sole reclamation agent or exploits the synergy of salt tolerant crop varieties with reduced quantities of amendments. While in the first case chemical amendment is fully substituted, in the later case only a part of it is substituted to reduce investments on this input. As such, biological reclamation approaches are most suited to resource poor farmers who are unable to invest on recommended doses of the chemical amendments.

Table 4.20 Credit worthiness of different components of an integrated farming system

Cropping system	Cost of cultivation (Rs.)	Gross income (Rs.)	Net returns (Rs.)	B:C ratio
Rice-wheat	5235	10103	4868	1.93
Forage	1198	2228	1030	1.85
Vegetable	2989	7472	4483	2.50
Fruit	3600	2100$	---	1.70$
Aquaculture	16254	42840	26586	2.63
Total	29276	64743	36967	2.21

$ The profits would increase with time

Highly alkali soils (pH > 10.0)

In the first approach, such soils could be reclaimed by ponding water and growing certain salt tolerant grasses. While Karnal grass proliferates under high ESP and initiates the reclamation process, para grass is also quite tolerant and could also be grown to reclaim alkali lands. The results revealed that after growing Karnal and para grasses for 1 year, satisfactory yield of rice could be obtained. After 2 years, even good yield of wheat could be obtained. During these years, good yield of Karnal and para grasses are also obtained and the products can be used as fodder (Table 4.21). Although slow and less efficient, it could be adopted by resource poor farmers who cannot afford high investments on reclamation program.

Table 4.21 Yield of Karnal and para grasses grown in alkali soils without any amendment

Grass	Year 1	Year 2	Year 3
Karnal	26.9	28.4	25.4
Para	21.9	29.3	31.5

The part data reproduced from Kumar and Abrol (1986)

In the second approach to reclaim such soils, synergy of the amendments and the salt tolerant crop varieties is utilized. In this approach, one could grow CSR10 variety of rice during *kharif* and sugar beet during the *rabi* season (rice-sugar beet cropping sequence) for three years with small doses of any amendment or even without it. The cropping of the land initiates a self reclamation process through dissolution of native $CaCO_3$ by biological action of plant roots. Within three years, the soil pH will come down from 10.0 to 9.3. In the fourth year, any other high yielding variety of rice in *kharif* and wheat (KRL 1-4) or raya CS 52 can be grown successfully.

Salt tolerant varieties of rice, wheat and mustard could also be grown with a small dose of about 50% of the recommended dose of the amendment. The data in Table 4.22 reveal that salt tolerant varieties gave higher yields at low doses of gypsum while normal high yielding varieties yielded higher at higher doses of amendments. The long-term results revealed that salt tolerant rice varieties could be grown up to two years with 25% GR. Later on these varieties could be replaced by other high yielding varieties. The replacement of salt tolerant wheat varieties with high yielding varieties could be made after three years in 25% GR treated plots. The economic analysis indicated that highest B:C ratio of 1.70 was obtained with 25% GR and salt tolerant varieties of rice and wheat.

Table 4.22 Effect of gypsum levels and varieties on grain yield (t/ha) of rice and wheat

Gypsum levels (% GR)	Rice			Wheat		
	CSR 13	**Pant 4**	**Mean**	**KRL 19**	**PBW 343**	**Mean**
0	1.03	0.64	0.83	0.64	0.50	0.57
15	3.48	2.93	3.20	1.90	1.62	1.76
25	4.71	4.56	4.63	3.03	3.18	3.15
50	4.97	5.02	4.96	3.46	3.67	3.51
Mean	3.54	3.29		2.28	2.22	

Moderately alkali soils (pH 9.2 to 9.8)

Rice CSR 10 is grown during *kharif* season as the first crop without application of soil amendment. After the rice crop, the soil pH will come down to a level so that wheat KRL 1–4 or Indian mustard varieties such as Pusa bold, Kranti, Varuna or CS 52 can be grown during the *rabi* season.

Gurusikaran Model

The model has been extensively and successfully tried and tested in Aligarh (Uttar Pradesh) by AFPRO, New Delhi. The basic components of this model are similar to the chemical amendment model. However, the crop diversification in this model begins from the second year in favor of fodder crops to allow the farmers to venture into animal husbandry. The diversification to animal husbandry helps to supplement the meager income of the farmers from arable cropping of the reclaimed lands. This model is recommended for farmers with small parcels of alkali lands (Tyagi *et al.*, 1996).

Model for Alkali Vertisols

This model has been developed to reclaim heavy textured black soils under rainfed conditions. The chemical amendment is applied as per the GR calculated for the soil. In this model, major intervention is land modification that improves surface drainage and supplements limited rainwater to cultivate rice in a limited area. Any upland crop such as maize or cotton is grown on beds (raised beds) and rice is grown in furrows (sunken beds) (Fig. 4.11). While such an arrangement helps to avoid submergence of maize and cotton, the drainage water from the beds supplements the rainwater for rice crop grown in furrows. The model has been applied to alkali Vertisols in the states of Madhya Pradesh, Gujarat and Maharashtra.

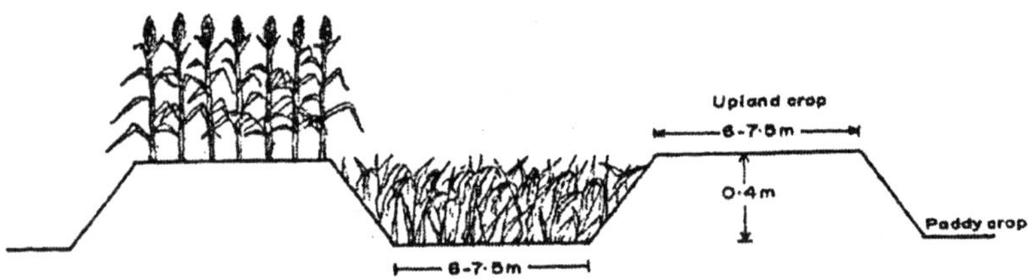

Fig. 4.11 A model to reclaim black cotton soils (Vertisols)

THERMAL HEATING AND RECLAMATION

Investigations have shown that heating an alkali soil to a high temperature can result in significant improvement in the properties of an alkali soil (Gupta and Bandyopadhya, 1975). On heating an alkali soil containing illite possibly vermiculite to 500°C for 2 h in a muffle furnace, the EC_e dropped from 13.4 to 4.8 dS/m, ESP from 92 to 37, boron from 9 to 1 mg/L and hydraulic conductivity increased from 0.2 to 1.0 cm/hr.

Germination percentage of wheat, which was nil initially increased to 100% in the heated soil. According to Gupta and Bandyopadhyay (1975), it could possibly be due to the fact that a part of the sodium ions on the exchange complex penetrates into the vacant octahedral positions in the dioctahedral lattice of the illite possibly vermiculite clay mineral. After entering the vacant octahedral positions, the sodium is rendered difficult to exchange or is fixed in a non-exchangeable from. The collapse of the interlayer space of the crystal lattice follows the loss of the swelling potential of the soil resulting in increased hydraulic conductivity. Boron might have reduced as a result of volatilization losses or due to burning of organic matter. Significant improvements are observed only in illite clay mineral with almost no change in the kaolinite type of clay minerals and only minor improvement in bentonite type of clay minerals. Possibly these results need to be tested at field scale and some mechanisms developed to heat the soil say by slowly moving a down draft furnace with its flame directed towards the ground. Harnessing solar energy may be another way to reduce the cost of this technology.

ALTERNATE LAND MANAGEMENT

A sizeable area of alkali lands is categorized as village community lands, Govt. lands reserved for specific purposes, lands lying abandoned along the canals, railway tracks and roads and absentee lands. Reclamation of such areas for growing annual crops like rice and wheat may be difficult because of common property rights. Raising grasses, trees and trees plus grasses on such lands appears promising options to produce food, fuel and forage on one hand and to increase forest cover, moderate climate change and other issues of environmental conservation on the other. During last about four decades, few viable alternate management options/technologies have been developed which can be applied in different agro-eco regions of the country albeit some location specific modifications. The package of practices includes identification of suitable grass, forest and fruit trees, planting techniques, filling mixture composition and selection of a suitable silvi-pasture/agroforestry model. The alternate models need to be evaluated for their efficiency for soil amelioration, their benefit cost ratio and environmental impacts.

Promising Forage Crops and Grasses

In general grasses are more tolerant to salt than most other crops. The grasses such as Karnal grass, Rhodes grass, Gatton panic and Bermuda grass can be grown without application of gypsum even up to pH 10.4. The amelioration potential of some of the grasses has already been discussed in a previous section of this chapter. The relative tolerance of grasses and forage crops to different levels of soil sodicity is given in Table 4.23. Most promising *kharif* forage crop is sorghum. However, during *rabi*, oat, *shaftal* and *berseem* can be grown successfully. The forage crops can also be grown using alkali/sodic water with appropriate neutralization of the residual sodium carbonate. Depending upon the severity of alkalinity, gypsum@ 2-4 t/ha is recommended for sorghum, sudan grass, *shaftal, berseem* and oats.

Table 4.23 Promising forage crops for alkali soils

pH*	Grass/forage crop (Botanical name)
Grasses	
10.4	Karnal grass (*Leptochloa fusca*), Rhodes grass (*Chloris gayana*), Gatton panic (*Panicum maximum*), Bermuda/Dub grass (*Cynodon dactylon*)
10.0	Para grass (*Brachiaria mutica*)
9.6	Blue panic (*Panicum antidotale*)
9.2	Hybrid Napier (*Pennisetum purpureum*), *Setaria grass* (*Setaria anceps*)
8.8	Anjan grass (*Cenchrus ciliaris*), Guinea grass (*Panicum maximum*) Deenanath grass (*Pennisetum pedicellatum*)
Forage crops (Rainy season, *kharif*)	
9.6	Sorghum (*Sorghum bicolor*), Sudan grass (*Sorghum sudanence*) *Dhaincha* (*Sesbania aculeate*)
9.2	Teosinte (*Euchlaena americanum*)
8.8	Cowpea (*Vigna unguiculata*), Guar (*Cyamopsis tetragonalob*)
Forage crops (Winter season, *rabi*)	
9.6	Oat (*Avena sativa*), Barley (*Hordeum vulgare*), *Shaftal*/Persian Clover (*Trifolium resupinatum*)
9.2	*Berseem* (*Trifolium alexandrinum*)

* The pH values at which yield reduction was 50% or less.

Salt Tolerant Forest and Fruit Trees

A list of the promising species of forest and fruit trees is given in Table 4.24 (Singh *et al.*, 1997; Dagar *et al.*, 2001) The most promising species for planting in soils having pH more than 10.0 included: *Pahari Kiker* (*Prosopis juliflora*), *babul* (*Acacia nilotica*), *Frans* (*Tamarix articulata*) and Australian pine (*Casuarina equisetifolia*). The most promising fruit species are: *Aonla* (*Emblica officinalis*), *Karaunda* (*Carissa carandus*) and Guava (*Psydium guajava*).

Table 4.24 Relative tolerance of tree species to soil alkalinity/ sodicity/ pH

Average pH_2 (0-120 cm)	**Fuel-wood/Timber species#**	**Fruit trees**
More than 10.0	*Prosopis juliflora* (Pahari kiker) *Acacia nilotica* (Kikar), *Casuarina equisetifolia* (Australian pine), *Tamarix articulata* (Frans) *Casurina equietifolia* (Jor-tor)*	-
9.5 to 10.0	*Terminalia arjuna* (Arjan) *Pithecello biumdulce* (Jangli jalebi) *Eucalyptus tereticornis* (Safeda) *Albizzia lebbek* (Siris) *Tamarindus indica* (Imly) *Acacia catechu* (Khair) *Butea monosperma* (Dhak) *Cordia rothii* (Lasura) *Pongamia pinnata* (Papri, Karanj) *Parkinsonai aculeata*	*Achras japota* (sapota) *Carissa carandus* (Karaunda) *Psidium guajava* (Guava) *Emblica officinalis* (Aonla) *Zizyphus maurtiana* (Ber) *Syzgium cumini* (Jamun) *Phoenix dectylifera* (Khajoor) *Aegle marmelos* (Bael)
9.0 – 9.5	*Dalbergia sisoo* (Shisham) *Azadirachta indica* (Neem) *Grevillia robusta* (Silver Oak) *Sesbania sesban* (Dhaincha) *Kigelia pinnata* (Kigelia) *Melia azedarach* (Bakain) *Leucaena leucocephala* (Subabul) *Moringa olifera* (Saunjana) *Bambusa* spp. (Bamboo) *Morus alba* (Mulberry) *Albizia procera* (Siris) *Acacia auriculiform, Acacia senegal*	*Punica granatum* (Pomegranate) *Prunus persica* (Peach) *Pyrus communis* (Pear) *Vitis vinifera* (Grapes) *Mangifera indica* (Mango)
< 9.0	*Tectona grandis* (Teak) *Populus deltoides* (Poplar) *Ficus virens, Bombax ceiba* *Hardwickia binnata,* *Thespesia populnea,* *Bauhinia variegate,* *Lagerstroemia speciosa,* *Cadrella toona*	

Source: #Singh *et al.* (1977); *Not suitable for deep ground water areas

Planting Techniques and Performance

Amongst the four commonly used plantation techniques such as pit method, auger hole method, ridge plantation and trench plantation, auger hole technique seems to be quite promising to raise trees in highly alkali soils (Fig. 4.12). Auger holes of 20-25 cm diameter and 120-140 cm deep (depending upon the presence of calcium carbonate layer in the soil profile) are made by a specially designed auger mounted on a tractor. These auger holes are filled back with a mixture of original soil blended with 3 kg gypsum, 8-10 kg farm yard manure and 10 kg river sand before planting the trees. Total cost of planting with this technique is about Rs. 30,000/ha. The biomass production from selected trees 10 years after planting by auger hole technique in a soil of original pH_2 10.2, is given in Table 4.25 (Singh and Dagar, 1985, Singh and Dagar, 2005) .

Fig. 4.12 Tractor operated auger hole making arrangement (left) and a one year old plantation with auger hole technique (right)

Amelioration by Trees

Trees reclaim alkali soils by addition of leaf litter, root activity and through the dissolution of native calcium carbonate present in these soils. Ameliorative effects of five multipurpose tree species raised on an alkali soil of pH 10.4 are given in Table 4.26. Within 20 years, the pH of the soil reduced from the original 10.3 to around 9.0 and organic carbon increased around four fold except in *Eucalyptus tereticornis*. The reclamation was higher in *Prosopis juliflora* compared to other tree species. All kinds of agricultural crops can be grown in alkali soils reclaimed by trees.

Table 4.25 Growth performance and biomass production of salt tolerant trees in a soil of pH_2 10.2 after 10 years of planting

Growth parameters	Tree plantations			
	Prosopis juliflora	*Acacia nilotica*	*Casuarina equisetifolia*	*Eucalyptus tereticornis*
Height (m)	12.9	11.6	14.5	14.9
DSH (cm)	15.9	15.4	15.6	13.6
DBH (cm)	12.5	13.6	12.0	11.0
Bole weight (kg/tree)	112.6	85.4	84.2	65.6
Branches + Leaves weight (kg/tree)	43.2	43.8	28.4	23.5

DSH : Diameter at stump height (5 cm above the ground level)

DBH : Diameter at breast height (137 cm above the ground level)

Table 4.26 Ameliorating effects of mesquite and other tree plantations on an alkali soil

Species	Original		After 20 years	
	pH_2	Organic carbon (%)	pH_2	Organic carbon (%)
Eucalyptus tereticornis	10.3	0.12	9.8	0.33
Acacia nilotica	10.3	0.12	9.03	0.55
Albizia lebbek	10.3	0.12	8.67	0.47
Terminalia arjuna	10.3	0.12	8.15	0.47
Prosopis juliflora	10.3	0.12	8.03	0.58

Planting Technique for Fruit Trees

Fruit trees, in general, are comparatively less tolerant than most forest trees. The relative tolerance of fruit trees as assessed by Singh *et al.* (1994) is given in Table 4.27. A pit-cum-auger hole technique is used to successfully raise the fruit plantations. In this technique, pits of 45 cm x 45cm x 45 cm dimensions are made manually and in the centre of each pit, auger holes of 20-25 cm diameter and 120-140 cm deep are made to facilitate strong feeder and deep root systems. The dose of gypsum, FYM and river bed silt which is blended with the original alkali soils is nearly double of what is used to raise forest trees.

Table 4.27 Some suitable fruit species and their relative tolerance

Fruit species	Tolerance		Suitable varieties
	pH	ESP	
Aonla- Embilica officinalis	9.5	40	Chakaiya, NA-7, NA-6
Guava - *Psydium guajava*	9.2	30	Sardar (L-49)
Ber- Zizyphus mauritiana	9.5	45	Banarsi, Karaka, Ponda
Bael - Ziziphus mauritiana	9.0	30	Narendra-5, Narendra-9
Mahua - Madhuca latifolia	9.2	-	Narendra selection
Grape- *Vitis vinifera*	9.0	30	Beauty seedless, Perlette
Karaunda-Carissa carandus	9.2	35	Narendra selection

Comparative performance of eight fruit trees planted by two different techniques after 10 years of planting is given in Table 4.28. The average fruit yield from *Aonla*, *Karaunda*, Guava and *Ber* from 6 years onwards in an alkali soil of pH 10.0 is given in Table 4.29 (Singh *et al.*, 2008). The financial analysis revealed that the B:C (benefit to cost) ratio for *Aonla*, *Karaunda* and Guava was 2.79, 1.62 and 1.60, respectively

Table 4.28 Comparative growth of eight fruit species in a alkali soil

Fruit species	Planting technique			
	Pit (90cm x 90cm x 90cm)		Pit-cum-auger hole	
	Survival (%)	Height (m)	Survival (%)	Height (m)
Jamun (*Syzygium cumini*)	100	4.82	100	5.05
Imli (*Tamarindus indica*)	96	3.45	96	3.54
Karaunda (*Carissa carandus*)	96	1.65	100	1.96
Ber (*Zizyphus mauritiana*)	96	3.03	100	4.16
Guava (*Psydium guajava*)	91	3.03	100	3.15
Aonla (*Emblica officinalis*)	88	2.61	96	4.17
Mango (*Mangifera indica*)	30	1.33	55	2.12
Pomegranate (*Punica granatum*)	20	1.55	45	2.12
Mean	77	2.68	86	3.28

Fuel Wood-cum-Forage Production System

In this system, salt tolerant trees like *Pahari Kikar* (*Prosopis juliflora*) and Karnal grass (*Laptochloa fusca*) are grown together without any amendment application. In addition to firewood and forage production, the system reclaims alkali soils to the extent that normal crops such as oats, *senji, berseem* and *shaftal* etc. can be grown after 5-6 years. Large area of salt affected village community lands has been reclaimed by following this bio-amelioration approach. Almost all parts of *Prosopis juliflora* can be used for economic purposes like furniture, firewood and pods as animal and human food (Singh *et al.*, 1993).

Table 4.29 Yield and economic gains from promising fruit trees after 6 years of plantation

Years after planting	Fruit species							
	Aonla		*Karaunda*		Guava		*Ber*	
	Y	I	Y	I	Y	I	Y	I
6	2.2	9.8	1.4	5.7	1.5	5.6	1.2	2.2
7	2.4	14.7	1.7	7.4	1.8	7.0	1.4	3.0
8	2.7	19.6	2.0	9.5	1.9	8.3	1.5	3.7
9	3.2	26.9	2.1	10.6	2.1	11.0	2.1	6.3
10	-	-	2.5	14.0	2.3	12.4	2.2	7.8

Y : Yield, I : Income in thousand rupees/ha

5

MANAGEMENT OF INLAND SALINE SOILS

The osmotic potential of the soil solution is the primary salinity variable in a saline soil which together with matric stress impacts the crop response. Although the total stress given by the sum of matric and osmotic stresses may not be the only cause of low crop yields as other factors come into play and influence the yield. Central Soil Salinity Research Institute, Karnal developed a package of practices to reclaim inland waterlogged saline lands. To obtain high sustainable yields during the reclamation phase and beyond, the complete package which comprises of on-farm land development, surface and subsurface drainage, leaching, selection of crops and cropping sequences and improved cultural practices needs to be implemented. Generic solutions of at least two components of this package i.e. land drainage to control water table and leaching processes to reduce excess soluble salts have been discussed in the foregoing chapters. Therefore, these components are only briefly covered with appropriate case studies. A major discussion in this chapter is made on the management and crop production aspects for arable cropping. Since arable cropping might pose problems in many settings of inland saline soils, alternative management technologies to be adopted in such situations also form a part of this chapter. These interventions, in the absence of sufficient funds for land drainage, would help to obtain some economic gains.

FACTORS AFFECTING CROP PRODUCTION

In saline soils plant growth is adversely affected due to the following factors:

- Excess soluble salts present in the soil solution depress the water potential of the medium that restricts the water uptake by plant roots

- High concentrations of some specific ions can be toxic and can induce physiological disorders (e.g. Na^+, Cl^-, SO_4^{--}, Mg^{2+}, B)
- High salinity may affect several metabolic processes such as CO_2 assimilation, protein synthesis and respiration
- In waterlogged salt affected soils, some elements may be produced and accumulated to toxic levels
- Imbalance in the uptake of ions.

Usually, not one but a combination of the above factors would be in operation to affect crop production and productivity in waterlogged saline lands.

Nature of Salt Stress and Osmotic Potential

The osmotic potential of the soil solution is the primary salinity variable to which crops respond. The total soil moisture stress comprises of the matric potential and the osmotic potential. Because of the additive nature, the sum of the matric and osmotic potentials determines the total soil moisture stress. Whereas, osmotic potential predominates in saline soil low-tension treatments, matric potential predominates in non-saline high-tension treatments (Fig. 5.1). Unless the specific ion or nutritional imbalance is caused by salinity, the crop yields are influenced to equal degree under similar variations of the total soil moisture stress, be it due to soil salinity or due to delayed irrigation (Abrol *et al.*, 1988). Soil salinity produces symptoms such as stunting of plants, a lower shoot-root ratio, and smaller, darker green (bluish-green) leaves which closely resemble the symptoms caused by drought conditions. A school of thought has thus emerged, which considers that the salinity affects the plants due to disturbance in the plant-water balance. This situation is commonly referred as `physiological drought'.

Osmotic potential, σ_{fc} of the soil solution at field capacity, is related to the electrolyte conductivity of the saturation extract, EC_e at 25^0C and is given by:

$$\sigma_{fc} = -7.8\ EC_e - 0.046\ EC_e^2 \qquad (5.1)$$

This relation assumes that the water content at field capacity is equal to half the water content at saturation. The osmotic potential is expressed in J/kg and EC_e in dS/m. A simple relationship between the osmotic pressure in atmosphere (O.P.) and electrolyte conductivity of the saturation extract in dS/m at 25°C can be approximated by the simple relationship:

$$O.P. = 0.39\ (EC_e \times 10^3) \tag{5.2}$$

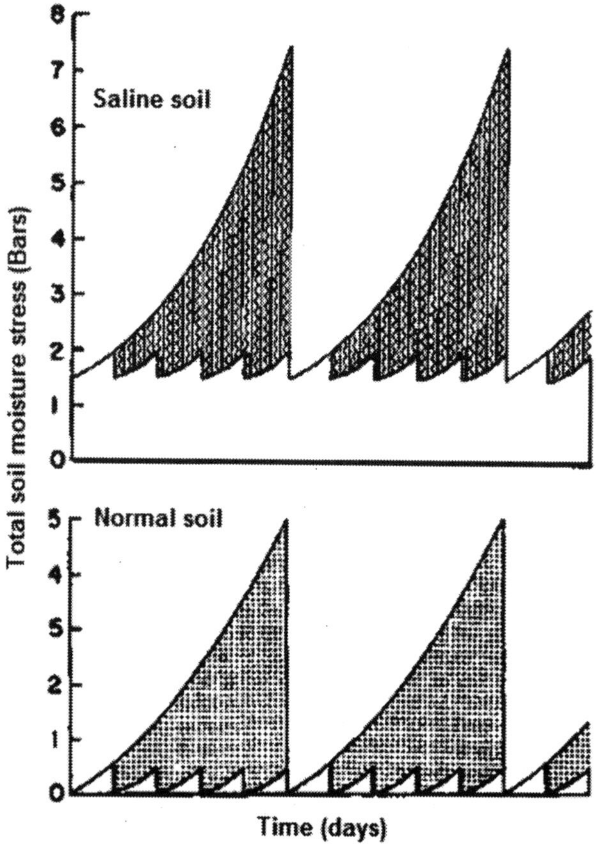

Fig. 5.1 Characteristics variations of moisture stress in irrigated soils (top) saline soils and (bottom) normal soil (Lower lines show frequently irrigated while upper lines show infrequently irrigated soils

RECLAMATION AND MANAGEMENT PACKAGE

- On-farm land development (grading, construction of field dykes (*bunds*) etc.)
- Surface and subsurface drainage
- Leaching
- Selection of crops and cropping sequences
- Improved management practices

ON-FARM LAND DEVELOPMENT

Amongst the several on-farm land development activities, construction of small dykes (*bunding*) and land leveling are essential to implement an effective land reclamation program. Strong *bunds* help to conserve rainfall within the plots to leach down the salts. In arid and semiarid regions, fresh water being limited, it helps to meet a part of the onetime leaching requirement of saline lands. Moreover, strong bunds also check the inflow of the surface runoff from un-reclaimed lands, which, if allowed to pass over the reclaimed lands, might reverse the reclamation process.

Proper land leveling is required to achieve uniform leaching. It also helps to apply shallow depths when surface irrigation systems are employed to irrigate the lands. Inappropriate land grading would not only delay the reclamation process but at times the whole program might collapse. The designed slope must fit in the grade requirement of a good surface irrigation and an efficient surface drainage. A grade in the range of 0.05 to 0.2% in the direction of drains is commonly used. The preferable slope is about 0.1%. In areas with flat topography, the grading operation should take advantage of the existing land slope. A field layout plan should be prepared considering irrigation and drainage requirement of the saline lands. Each field should be designed in such a way that it has an irrigation channel at the head end and a drainage channel at the tail end.

Scraping

In order to get quick returns, farmers try to remove the salts that gets accumulated on the soil surface using manual or mechanical means. They scrap about 5 cm of the soil and keep it as a heap in one corner of the field. Scientifically speaking, scarping might provide temporary relief for 1-2 years. The practice has limited utility especially for areas with shallow water table (Fig. 5.2). The reduced depth to water table accentuates the process of capillary action, increases the evaporation rate and causes quicker soil salinization. Ultimate disposal of salts would also continue to be a major problem. Therefore, scraping as a means of land reclamation does not find much favor.

Realizing these problems, a farmer in Palwal district in Haryana adopted a quite different approach. He raised the surface of his field by adding a layer of normal soil. Although the cost of such an operation is enormous yet it should obviate some of the problems encountered in scraping. The water table is now deeper by the amount the field level is raised the disposal related problem is altogether avoided (Fig. 5.2).

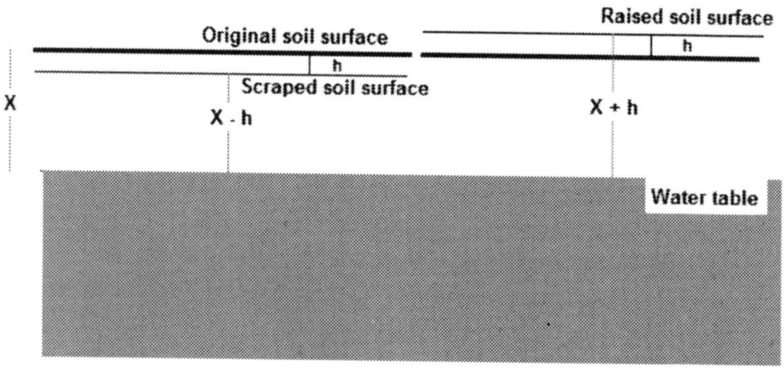

Fig. 5.2 Ground water depth in relation to scraping and raising the field surface

SURFACE AND SUBSURFACE DRAINAGE

Surface drainage should be an integral part of agriculture in monsoon climatic conditions. Even where subsurface drainage is provided, surface drainage should be integrated with it because the whole runoff cannot be drained through subsurface drainage. Besides posing few insurmountable problems such an attempt would enormously increase the cost of subsurface drainage. In surface drainage, ditches are provided to remove excess runoff that fails to enter the soil. Field ditches empty into collector ditches build to follow a natural water course. However, it may not be wise to drain the whole water quickly. Attempt should be made to store the water into the soil to help in the leaching process. The capacity of a drainage system is decided on the basis of the amount and frequency of heavy storms (Chapter 2). The area, a ditch can satisfactorily drain, depends on how quickly water runs into the ditch and the size of the ditch and its slope. Although several other layouts exist, yet field ditches are usually laid parallel to each other 20 to 60 m apart. These are 30 to 45 cm deep depending upon the depth of the collecting ditch.

Subsurface Drainage

The purpose of this section is to highlight that the generic solutions discussed in Chapter 2 have been and could be successfully utilized in reclamation projects. Work on subsurface drainage for reclamation of salt affected lands was initiated in the year 1980 by Central Soil Salinity Research Institute, Karnal at Sampla (Gupta, 1985). As soon as a subsurface drainage is installed and other management practices are imposed, leaching and desalinization processes become operative. Salt content of the root zone continues to reduce over the years as long as land reclamation process remains in operation (Table 5.1). Desalinization is quick in semiarid regions where rainfall is on an average 500 mm or more. In these areas, a salt tolerant crop can be cultivated even in the first year of reclamation. Any crop can be grown once the desalinization is complete in 2-4 years time. In the initial years of desalinization, crops like pearl millet, mustard, wheat and barley were successfully grown at sampla (Rao *et al.*, 1986). The yields of these crops continued to increase until yields were comparable to normal soils of the region (Table 5.2). Similar results were obtained in Karnataka and Andhra Pradesh where rice yield increased over the years due to desalinization of the root zone (Fig. 5.3).

Table 5.1 Effect of subsurface drainage on soil salinity (dS/m)

Soil depth	Drain spacing (m)											
	25				50				75			
	A	B	C	D	A	B	C	D	A	B	C	D
0-20	50.7	5.3	2.9	2.7	50.7	8.1	3.6	2.0	46.1	8.3	6.1	4.2
20-40	23.6	4.0	3.3	3.7	19.4	4.7	3.4	2.1	26.4	9.1	7.8	5.0
40-60	19.4	3.7	3.7	1.9	15.8	7.9	4.3	3.1	13.4	9.0	9.4	7.7
60-90	17.0	4.8	5.0	2.0	16.8	11.1	5.8	5.0	11.1	9.4	8.7	7.1
90-120	12.2	7.6	6.2	2.4	15.5	14.3	7.1	4.4	12.6	10.2	10.7	8.4
120-150	-	-	6.0	2.6	-	-	9.3	4.6	-	-	12.5	10.0
150-200	-	-	6.6	3.1	-	-	9.9	6.8	-	-	12.1	10.4

A=June 1984, B=Oct 1984, C=Nov. 1984, D=Oct. 1988

Table 5.2 Yield of different crops (t/ha) as affected by subsurface drainage at Sampla

Crop	1984-85			1987-88		
	25	50	75	25	50	75
Wheat	4.9	4.0	2.5	5.1	5.4	4.9
Barley	4.2	2.9	2.0	2.9	3.1	2.2
Mustard	2.4	2.0	0.9	-	-	-

Considering the success of these experiments, pilot and large scale subsurface drainage experiments were initiated to test the technology in Haryana to develop design criteria. The following guidelines are now followed in land reclamation programs of the state.

Drain depth	Shallow drains (< 2.0 m) preferably around 1.5 m for pumped outlets
Drainage coefficient	1.5-2.5 mm/d
Drain spacing	75 m drain spacing for medium texture soils of Sampla, later modified in HOPP as 60-67 m to be on *killa* lines
Pipes	PVC corrugated pipes
Envelope requirement	Geosynthetic envelop with O_{90} >300 on laterals

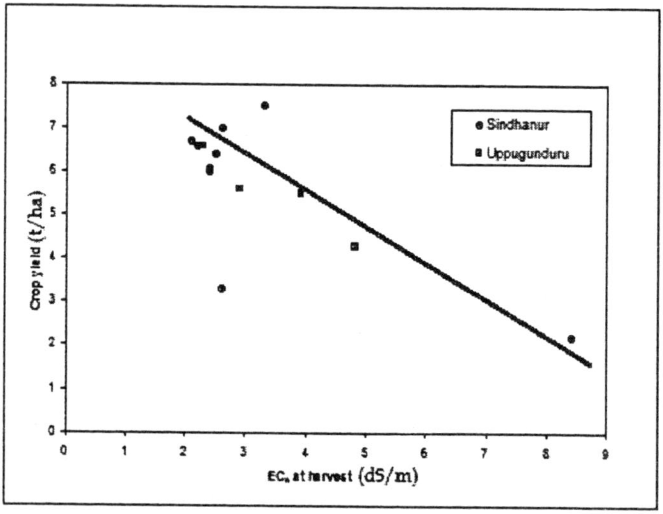

Fig. 5.3 Rice yield as affected by desalinization following land drainage

The technology has been extensively tested in other states as well. Some large scale projects were taken up in Kota, Rajasthan (RAJAD, 1995), Gohana (Sharma and Gupta, 2006) and Kalayat in Haryana and Sangli in Maharashtra (Dandekar and Chougule, 2010). RAJAD was a large scale applied research project on the use of horizontal subsurface drainage (SSD) and associated water management techniques to control soil salinity and water logging problems in irrigated agricultural lands in Chambal Command in Rajasthan. It was supported by Canadian International Development Agency (CIDA). About 15,000 ha were put under SSD including test plots of 1,400 ha (Table 5.3). Based on test plot results from 1400 ha, the following design criteria were developed and adopted (Tejawat, 2007):

Drain depth	1.2 m
Drainage coefficient	1.5 mm/d for laterals and 3 mm/d for collectors to allow for additional seepage from canals and surface drains
Drain spacing	35–60 m to provide satisfactory salinity and water table control
Envelope requirement	Not required if soil profile clay % is >40
	Required if clay % <30 or clay content is between 30 to 40% with soil sodicity, SAR >8

Avoiding details of individual experiments, it may suffice to list the various drainage technologies that have been tested in various states to reclaim waterlogged saline soils (Gupta, 2011; Table 5.4). The emerging guidelines on drain depth, drainage coefficients and lateral drain spacing are listed as follows.

Drain depth	1.5 m	Pumped outlets
	1.1 m	Gravity outlets
Drainage coefficient	1 mm/d	Arid regions
	2 mm/d	Semiarid regions
	3 mm/d	Humid regions
Drain spacing	100-150 m	Light texture
	50-100 m	Medium texture
	30-50 m	Heavy texture

Table 5.3 SSD installations in RAJAD Project

Project	Area installed (ha)	Installation equipment	Year
1. Small test plots	410	Excavator/Backhoe	1991-1993
2 Large test plots	1010	Integrated trenchless plow	1993-94
3 Pre CON/1 (Training phase)	700	Trenchless plow (Bull dozer with plow attachment)	1995-96
4. CON/1	10,671	Mastenbroek Trencher, V-Plow and Wolf Plow	1996-99
5. Areas with patchy salinity	2,134	Excavator/Backhoe	1997-99
Total	14925		1991-99

Source: Ram *et al.* (2000)

LEACHING

Subsurface drainage alone is not an end of a reclamation program. Unless the excess salts are leached out of the root zone, it would be difficult to grow a crop or improve the land productivity. Any of the leaching methods discussed in Chapter 3 such as flushing, continuous ponding or intermittent ponding could be used. Sometimes one may need to apply a chemical amendment to facilitate leaching especially to soils having high SAR (Khosla *et al.,* 1979). Besides, rainwater management and use of saline drainage effluents to leach soils could minimize the dependence on fresh water. The leaching methodology of subsurface drained lands could be slightly different than the generic solutions. As such, it is being discussed in the following section.

Table 5.4 Drainage technologies tested in different states of India

State	Drainage technologies tested
Andhra Pradesh	Shallow pipe drains, deep open drains (Gravity/pumped outlets)
Assam	Pipe drains, open drains for tea gardens
Gujarat	Shallow pipe and deep open drains, vertical drainage+
Haryana	Deep/shallow pipe drains*, deep open drains^, open well drainage, bio-drainage (Pumped outlets, evaporation ponds)
Karnataka	Shallow pipe drains, bio-interceptor drains (gravity outlets)
Kerala	Shallow pipe drains for acid sulphate soils
Madhya Pradesh	Shallow pipe drains, mole drains
Punjab	Shallow pipe drains, vertical drainage, skimming well drainage, bio-drainage
Rajasthan	Shallow pipe drainage, open well drainage+, Open interceptor drains, Deep open seepage drains (Pumped outlets, evaporation tanks)
Uttar Pradesh	Pipe and bio-interceptor drains
Uttrakhand	Shallow pipe drainage for *terai* region

*deep pipe drains > 2 m depth; shallow pipe drains < 2 m depth; ^ open drains deeper than 0.9 m; + Results were not encouraging; Text in brackets indicate technologies for drainage effluent disposal

Leaching Drained Lands

To achieve uniform leaching in subsurface drained lands is an uphill task. If water is ponded on the entire field between the two lateral lines, more water flows from areas nearer to the drains than from the area midway between the drains. Water flow is theoretically zero at the center point between the two laterals. Thus, areas nearer to the drains will be leached more than the areas at the center of the drains. The non uniformity of leaching is more when drains are spaced farther apart than when closely spaced. Gupta (1983) observed non uniformity of leaching in lands when drains were spaced 50 m apart but yields were more or less at par with distance from the drains. It was mainly because the non uniformity was confined to depths at or below 30 cm from the ground level. In another trial, yield of pearl millet varied with distance from the drains at Sampla in the first year of cultivation in 50 and 75 m spaced drains. The yield was high up to 20 m away from the drains and declined thereafter. Laboratory tests proved that partial ponding of the fields could to a greater extent take care of such situations and improve uniformity of leaching (Batta *et al.*, 1980). Thus, prior to leaching, smaller plots with temporary *bunds* within the designed layout are constructed. These bunds run parallel to the drains (Fig. 5.4). Note that in this figure only half of the field between the two laterals is shown. The other half would be a mirror image of the same. Rao and Leeds-Harrison (1990) conducted a simulation study and concluded that uniform leaching could be obtained by ponding the entire plot for 12.5 days followed by ponding the mid-half of the plot for 12.5 days and ponding of the central quarter of the plot for another 12.5 days (Fig. 5.4). Although, it takes more time, yet such a scheme results in complete and uniform leaching of the drained land. A strategy like this also saves about 50% water compared to the amount of water required to leach the field to the same level by complete ponding of the plot for the whole duration. The duration of ponding could be increased or decreased depending upon the soil type, initial salinity of the soil, desired salinity of the soil and the water availability.

SELECTION OF CROPS AND CROPPNG SEQUENCES

Selection of crops and cropping sequences for saline soils is of paramount importance, since crops differ widely in their tolerance to salinity. The adverse effects of salt stress on plant growth could be minimized to some extent by growing salt tolerant crops or their varieties. Qualitatively crops are categorized in 4 groups depending upon their tolerance to soil salinity (Table 5.5). It should be noted that these rating are only guidelines for the relative salt tolerance among crops and minor variations might occur between tables given by various workers.

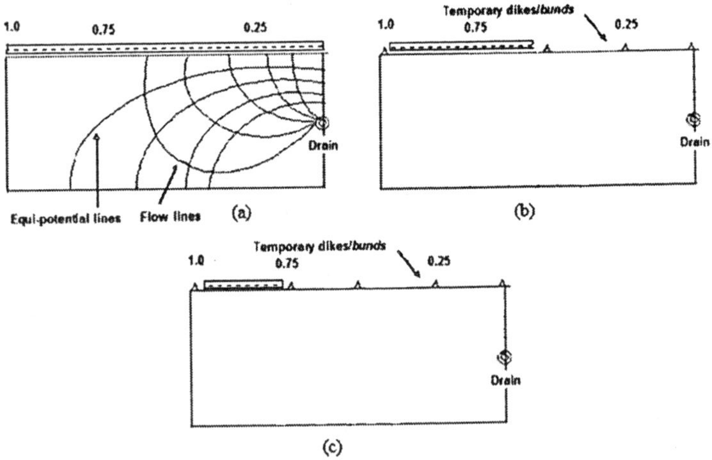

Fig. 5.4 Leaching strategy for the drained lands (a) flow lines and equi-potential lines with fully ponded field (b) Ponding central half and (c) ponding central quarter of the field

Table 5.5 Crop groups based on response to salt stress

Sensitive groups		**Resistant groups**	
Highly sensitive	**Medium sensitive**	**Medium tolerant**	**Highly tolerant**
Lentil	Radish	Spinach	Barley
Mash	Cowpea	Sugarcane	Rice (transplanted)
Chickpea	Broad bean	Indian mustard	Cotton
Beans	Vetch	Rice (direct sowing)	Sugar beet
Peas	Cabbage	Wheat	Turnip
Carrot	Cauliflower	Pearl millet	Tobacco
Onion	Cucumber	Oats	Safflower
Lemon	Gourds	Alfalfa	*Taramira*
Orange	Tomato	Blue panic grass	Karnal grass
Grape	Sweet potato	Para grass	Date palm
Peach	Sorghum	Rhodes grass	*Ber*
Plum	Minor millets	Sudan grass	Mesquite
Pear	Maize	Guava	*Casurina*
Apple	Clover, *berseem*	Pomegranate *Acacia*	*Tamarix* *Salvadora*

Selection of the first crop during reclamation will depend upon the degree of soil salinity after the basic reclamation measures such as land drainage, field leveling, bunding and leaching have been implemented. Since complete reclamation cannot be attained in the initial years, tolerant crops that can cope up with the salinity levels are preferred in the early phases (Table 5.5). Cultivation of crops having low evapotranspiration and/or high tolerance is one way of compensating for water deficiency. It would be prudent to know that crops also differ in their salt tolerance at various growth stages (Table 5.6). Additional information on this issue for Indian conditions is given in Table 5.7. Therefore, while selecting crops and irrigation strategy especially irrigation with saline water, this variation must be kept in view.

Table 5.6 Tolerance of various crops to salinity at two growth stages

Crop	Germination stage	Established stage
Barley	Very good	Good
Corn (maize)	Good	Poor
Wheat	Fairly good	Fair
Alfalfa	Poor	Good
Sugar beet	Very poor	Good
Beans	Very poor	Very poor

Source: Canada Department of Agriculture (1977)

Improved Varieties

Not only crops but even the crop varieties are known to differ in tolerance to soil salinity. Improved varieties, released by the Central and State Varietal Release Committees for a particular agro climatic region should be used for commercial cultivation. Improved varieties of different crops released for cultivation for northwestern zone are given in Table 5.8.

Criteria for Rating Crop Salt Tolerance

The ability of crops to tolerate salinity is determined by measuring one or more of the following four parameters:

- Germination
- Plant survival
- Absolute plant growth or yield
- Relative growth or yield

Table 5.7 Relative sensitivity of the crops at different growth stages

Crop	Relative sensitivity
Agra, Sandy loam soil	
Mustard	Pre-sowing (12.6) > flower initiation (16.9) > secondary branching (28.0)
Wheat	Pre-sowing (9.0) > flowering (13.3) > milking (20.0) > crown root initiation (24.0) > jointing
Barley	Crown root initiation (15.9) > pre-sowing (23.9) > flowering/booting (23.9) > jointing
Safflower	Pre-sowing (8.1) = Rosette (8.1) > flower initiation (12.1) > main head opening (24.2)
Dharwad, silty clay loam, black soil	
Paddy	Transplanting (2.7) > 50% flowering (6.0) > end of tillering (7.3)
Wheat	Pre-sowing (2.9) > crown root initiation (5.7) > milking (6.7) > flowering (8.9) > jointing (9.2)
Maize	Silking (6.9) > tasseling (7.6) > pre sowing (14.1) > knee height
Pigeon pea	Pre-sowing (2.1) > flowering to pod development (4.2) > flowering (5.7) > seedling to flowering (7.1)
Bapatla, silty clay loam and loamy sand soil	
Sunflower	Seeding (2.0) > flower bud initiation (4.7) > grain filling (4.8) > germination to flowering (6.2)
Onion	Transplanting to bud formation (1.6) > Bulb formation to bulb development (14.1) > bulb development to maturity (5.9)
Groundnut	Germination to pegging (2.3) > pegging to pod formation (4.5) > pod formation to maturity (5.9)
Black gram	Flowering to maturity (1.1) > seeding (3.2) > seeding to flowering (6.1)

EC_{iw} values in parentheses indicate 10% reduction in potential yield, if water of the indicated EC_{iw} (dS/m) is used at the specified growth stage and non-saline water is used at all other stages

Large number of investigations on differential response of crops or varieties have been reported at germination stage and attempts made to extrapolate the tolerance limit for final performance. More often than not, such attempts are likely to be frustrating because tolerance characteristics differ from one stage of growth to another.

Plant survival on saline soils has been widely used by ecologists to judge the tolerance of natural species growing on uncultivated lands. To the farmer, survival without an economic yield obviously has little value. Indirectly, this criterion is potentially useful for plant breeders to use it as the basis for selecting salt tolerant lines and cultivars. The only agronomical important plant response measurement to establish salt tolerance of a crop is the commercial yield. Absolute yields are certainly of greatest interest to the farmers. Absolute salt tolerance is directly related to the maximum growth or yield potential of a given plant under high salinity regardless of its growth rate under non-saline conditions. Under salinity stress, plant growth may decrease substantially without affecting its ranking for salt tolerance. For crops that have use both in its vegetative form and in the seed *e.g.* alfalfa, pearl millet, corn etc., the response to salinity should be assessed on both accounts. Too often vegetative response alone is unreliable to predict salt effects on fruit or seed production (Maas and Hoffman, 1977). Salinity may decrease grain yields of corn and rice without appreciably affecting straw yields.

Table 5.8 Improved varieties of different crops for saline soils of the northwestern plains

Crop	Variety
Wheat	
Irrigated timely sown	HD2009, HD 2329, HD 2428, PBW 154, Raj 1972, KRL I-4, and WH 157
Irrigated late sown	HD 2285, HD 2270, PBW 226, WH 2915 and Raj 2184
Barley	Ratna, Jyoti, Amber, DL 88, BL 2, DL 200, K 125 and RD 2423.
Mustard	Pusa Bold, Kranti, Varuna, Pusa *Barani*, Krishna, Prakash, RH 30, PR 45, Vaibhav, Vardan and Rohini
Pearl millet	PHB 10, MBH 110, HHB 60, MIH 179, CM 46, PSB 8, ICMS 7704 and WC C-75
Cotton	Laxmi, Sharda, H 777, HS 6, Pusa 31, F 414, F286, LH 886, LH 900, Bikaneri Nerma, G 1, G 27, AK235, AKH 4, Sangam, Amravati and MCU 7

Source: Minhas and Gupta, 1992

Relative yield has been defined as the yield of a crop grown on a saline soil expressed as a fraction of that would be obtained under non-saline conditions.

$$\text{Relative yield} = \frac{\text{Yield in saline environment}}{\text{Yield in non-saline environment}} \qquad (5.3)$$

It is expressed as a fraction or as a percentage by multiplying it with 100. It is assumed that all other factors that affect crop production are similar under the two situations. Thus, the reliability of relative yield data depends upon whether yield reductions are unaffected by the level of other essential nutrients. This criterion also permits a comparison between crops whose yields are expressed in quite different units. Relative tolerance also contributes to the overall fitness of the species to produce satisfactorily in a range of saline and non-saline environments, a character referred to as environmental plasticity.

Salt Tolerance Function

According to Maas and Hoffman (1977), the relative yield of a crop could be expressed in the form of a piece-wise linear regression equation and is described as:

$$Y = \frac{100}{0 \le EC \le EC_t} \tag{5.4}$$

$$Y = \frac{100 - S(EC - EC_t)}{EC_t \le EC \le EC_0} \tag{5.5}$$

$$Y = \frac{0}{EC \ge EC_0} \tag{5.6}$$

Here Y is the relative yield of the crop, EC is the electrolyte conductivity of the saturation extract, EC_t is the threshold EC up to which there is no decline in the yield of the crop, S is the slope and represents the yield reduction per unit increase in the EC of the soil solution and EC_0 is the EC at or beyond which crop fails to give any yield. A general plot of the model indicating yield variation with soil salinity is shown in Fig. 5.5. Although the model is known to have some limitations in describing salinity response functions for some crops, nevertheless it has been widely tested and used. Recently, this model has been validated to describe crop response to specific ion toxicity due to boron, exchangeable sodium percentage (Gupta and Sharma, 1990) and for salinity of the irrigation water (Misra *et al.*, 1990).

On the basis of this model, salt tolerance data could be described in terms of two key parameters:

(1) Threshold salinity, i.e. the maximum soil salinity that does not reduce yield below that produced under non-saline conditions; and

(2) Yield reduction per unit salinity greater than the threshold.

With these two parameters, the relative yield at any given salinity (EC_e) exceeding the threshold can be calculated by the following equation:

$$Y = 100 - S\,(EC_e - EC_t) \tag{5.7}$$

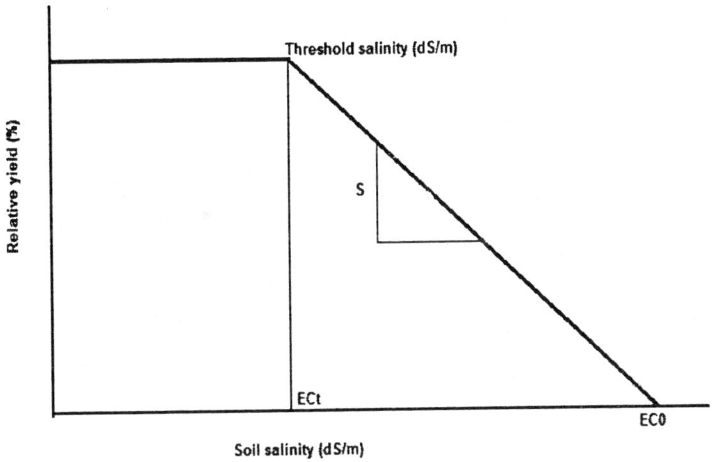

Fig. 5.5 Crop response function based on piecewise linear model

On this basis, crop response functions for various crops under different agro-climatic conditions have been derived and relevant parameters presented in Table 5.9 (Sharma *et al.*, 1992). More comprehensive data have been published by Maas and Hoffman (1977) and Mass (1986). The data of Maas (1986) is reproduced in Table 5.10. Any rating, however, will not provide an accurate estimate of the expected yield as the yield depends not only upon level of salinity but also upon many other cultural and environmental factors. The interaction between salinity and various soil, water and climatic factors modifies the ability of the plant to tolerate salinity. Pearl millet and cotton can be grown in soils up to an EC_e of 6.5 and 7.5 dS/m, respectively without significant yield reduction (Singh *et al.*, 1992).

Toxicity and Plant Tolerance

The specific ions affect the plants in numerous ways and the actual mechanisms in many cases are not well understood. Specific ions may influence respiratory pathways as in pea roots or inhibit the production of chlorophyll and carotene as in grapefruit. The injury due to specific ions depends upon the type of the salt and the plant species or even the plant varieties. Most chloride sensitive plants may be injured when Cl in soil saturation extract exceeds 5 to 10 meq/L. The chloride tolerant plants, however, can tolerate Cl concentration up to 30 meq/L. Excess sodium can cause nutritional disorders by reducing uptake of calcium and magnesium. Sodium, in addition to affecting plant nutrition, is known to adversely affect the soil physical conditions imparting low permeability to water and air and accumulation of black solubilized organic matter on the soil surface. This limits water and air infiltration and root penetration. Out of three commonly occurring toxic elements, B is most toxic and F the least. Critical limits could be considered as 5, 15 and 25 ppm for B, Li and F, respectively for tolerant crops like wheat (Gupta, 1984 b).

Table 5.9 Crop tolerance to soil salinity

Sites	Soil type	Crop	Threshold EC_t (dS/m)	Slope (%)	EC_{50} (dS/m)
Sampla	Sandy loam	Wheat	4.0	29.0	5.7
		Mustard	6.0	15.0	9.3
		Barley	7.0	19.0	9.6
Karnal	Sandy loam	Moong bean	1.8	20.7	3.3
		Mustard	3.8	6.9	11.0
		Sorghum	2.2	10.6	6.9
Agra	Sandy loam	Wheat	8.2	19.8	10.7
		Mustard	6.1	20.7	8.5
		Berseem	3.5	12.5	7.5
		Tomato	1.3	6.5	9.0
Dharwad	Black clay	Wheat	2.3	20.5	4.7
		Safflower	2.8	20.7	5.2
		Sorghum (w)	2.1	3.9	14.9
		Italian millet	6.1	50.0	7.1
Indore	Black clay	*Berseem*	2.3	20.5	4.7
		Safflower	2.8	20.7	5.2
		Sorghum (w)	2.1	3.9	14.9
		Italian millet	6.1	50.0	7.1
		Sataria	1.1	2.2	23.9

Cropping Sequence

In saline soils of arid and semiarid areas of northwest India, cotton, sorghum, pearl millet, cluster bean and *moth* bean could be grown during summer. During winter, wheat, barley, mustard, safflower and sugarcane could be grown depending upon the availability of water. The recommended cropping sequences for saline soils are pearl millet-barley, pearl millet-wheat, pearl millet-mustard, sorghum-wheat or barley, sorghum- mustard, cluster bean-wheat or barley and cotton-wheat or barley. Pearl millet-wheat, pearl millet-barley, pearl millet-mustard, sorghum (fodder)-wheat and sorghum (fodder)-mustard cropping sequences are more remunerative in saline soils (Singh *et al.*, 1992). Cotton based cropping sequences are not beneficial since the yield of the winter crops following cotton are low. In areas with water scarcity, mustard could replace wheat in the cropping sequence since its water requirement is less than wheat.

Table 5.10 Salt tolerance of some important agricultural crops (Mass, 1986)

Salt tolerance	EC values	Crops
Sensitive	$EC_t < 1.2$ $EC_0 < 8$	Chickpea, Beans, Sesame, Onion, Carrot, Green gram
Moderately sensitive	EC_t 1.2- 3 EC_0 8 - 16	Broad bean, Maize, Groundnut, Sugarcane, Alfalfa, *Berseem*, Radish, Paddy, Cauliflower, Chilly
Moderately tolerant	EC_t 3 - 6 EC_0 16 - 24	Wheat, Mustard, Cowpea, Oats, Sorghum, Soybean, Spinach, Pearl millet, Muskmelon
Tolerant	EC_t 6 -10 EC_0 24- 32	Barley, Cotton, Sugar beet, Safflower

MANAGEMENT PRACTICES

To establish a good crop stand in saline soils is a challenging task. Unlike normal soils, cultural practices for crop production in saline soils are quite different. An ideal package of cultural practices besides soil fertility and irrigation water management can ensure a good crop stand *vis-à-vis* good yield.

Seed Rate and Seed Treatment

Because of the inhospitable soil environment, germination in saline soils is inhibited. Mortality of young seedlings is common in saline soils. Coupled with poor tillering, crops yields are generally poor. Thus, to counter these effects a high seed rate is used (Table 5.11). For cotton, pearl millet, sorghum, wheat, barley, mustard etc. about 25% higher seed rate over the recommended seed rate for normal soils ensures good crop stand. In case of transplanted crops, the number of seedlings per hill should be increased and/or plant to plant and row to row distance should be decreased.

Table 5.11 Recommended seed rate for different crops in saline soils

Crop	Sowing time		Seed rate (kg/ha) and sowing time			
	Normal	Late	Normal soil		Saline soil	
			Normal	Late	Normal	Late
Wheat	1 to III week of Nov.	IV week of Nov. to III week of Dec.	100	120	125	145
Barley	Middle of Oct. to mid of Nov.	Whole of Dec. to 1 week of Jan.	80	100	100	125
Mustard	I to III week of Oct.	IV week of Oct . to mid of Nov.	4	5	5	6
Cotton	I week of April to mid May	Mid to end of May	12 (*Desi*)	15	15	18
			20 (Am.)	25	25	30
Pearl millet	Last week of June to mid July	III week to end of July	4	5	5	6

Am = American

The seeds should be sown in lines at an optimum depth with bullock or tractor driven seed drills. Seed-cum-fertilizer drill should be used to ensure uniform placement of seeds and fertilizers at optimum depth. Sowing with seed drill will give a uniform stand and lead to early emergence of seedlings. If seed drill is not available, sowing should be done by *kera* or *pora* method. The method, spacing and depth of sowing of some important crops in saline soils are given in Table 5.12.

Chemical treatment of seeds improves salt tolerance of crops. In a field experiment on saline soil (pH 7.6, EC_e 10.50 dS/m) highest yield of wheat was obtained when seeds were treated with 3% Na_2SO_4 solution. The yield increased by 351.0 and 217.3% in wheat varieties S-227 and K-65 respectively as compared to control (Puntamkar *et al.*, 1971). In another study, it was observed that grain yields of wheat and barley were not significantly affected when seeds were pre-soaked in salt solution. However, pre-soaking in low salinity tube well water resulted in more grain yield than seeds pre-soaked in NaCl or Na_2SO_4 solutions.

Table 5.12 Method, spacing and depth of sowing of different crops

Crop	Method of sowing	Spacing	Depth of sowing (cm)
Wheat	Line sowing with seed-cum-fertilizer drill	20-22.5 cm between rows and 18-20 cm for late sown crop	5-6
Barley	Line sowing with seed-cum - fertilizer drill	20-22.5 cm between rows and 18-20 cm for late sown	3-5
Mustard	Line sowing (Ridge and furrow, the seed should be sown in furrow or on the north side of the ridges)	30-40 cm between lines, Thinning of plants 15-20 days after sowing is necessary to maintain 12-15 cm plant to plant spacing	3-4
Cotton	Ridge and furrow method of sowing is better in saline soils. Ridges are made about 20 cm high at the desired spacing. Cotton sown on the side of ridges in east west direction. Seeds should be sown on the ridges at a height of 10 cm from the base of the ridge	60-75 cm between rows and 30-35 cm between plants in rows for American cotton. For *Desi* Cotton, 45-60 cm between rows and 25-30 cm between plants. Extra plants should be removed two to three weeks after sowing to maintain the desired spacing between the plants	5-6
Pearl millet	Ridge and furrow method. Sowing in the furrow in east-west direction, keep the seed zone wet such that salts move to the top of the ridge (higher concentration on the southern side)	60 cm between lines and 15 cm between plants. On any rainy day when the crop is 3 weeks old, healthy plants could be uprooted from places of over-crowding for gap filling	2-3

Pre-soaking of seeds in 50, 100, 200 and 300 mg/L of GA, IAA, IBA and NAA on the growth and development of wheat (S-227) irrigated with saline water (EC_{iw} 12 dS/m, SAR 30, boron 2 and 4 mg/L) indicated that phytohormone caused differential effect in plant height, root length, shoot production and grain yield. In general, the hormones produced significant positive changes in various growth attributes up to 200 mg/L but affected these parameters adversely at 300 mg/L. The beneficial effects were more at higher levels. It seems that use of pre-soaked seeds in salt or hormone solutions could be a beneficial cultural/management practice. However, much more efforts are needed to firm up sound recommendations on this issue.

Soil Fertility Management

The salt affected soils often lack in most essential plant nutrients owing to sparse vegetation and low organic matter content. Nitrogen deficiency is widespread in saline soils and a large fraction of the applied nitrogen under high soil salinity is lost in gaseous forms. Moreover, salts accumulation in saline soils affects the nutrient availability to plants in following ways: by changing the forms in which the nutrients are present in soils; by increasing the losses through leaching when the saline soils are leached heavily or through denitrification in nitrogen or by precipitation in soils; poor mineralization of organic N to inorganic forms due to high salinity and low microbial activity, poor symbiotic nitrogen fixation of atmospheric nitrogen due to adverse effect of salinity on rhizobia and legumes, the host plants, through interactive effects of cations and anions and through the effects of complementary (non-nutrient) ions on nutrient uptake. As such, adverse effect of high salinity can be alleviated to some extent by application of additional doses of nitrogen and use of balanced fertilizers. Many workers have observed positive interaction between salinity and added plant nutrients thereby resulting in higher yield. Crop yield generally decrease with increasing salinity but for a given salinity level the yield is increased by judicious use of fertilizers. Crops invariably respond to applied N in saline soils being deficient in available nitrogen.

High salt concentration (KCl or K_2SO_4) inhibits nitrification and results in accumulation of NH_4-N. Hence, the plants which absorb N only as NO_3-N will show nitrogen deficiency even if nitrogen in other forms is present in the soil. Singh *et al.* (1992) reported a significant increase in wheat yield up to 160 kg N/ha in saline soils (Table 5.13). They also recorded significant response of N on barley, mustard, pearl millet and cotton up to 120, 100, 120 and 80 kg N/ha, respectively. These levels are nearly 25% higher than normal dose of nitrogen for these crops in non-saline soils. Under low to medium salinity, urea is better than calcium ammonium nitrate and ammonium sulphate. But at high salinity, it is better to use NO_3^- containing fertilizers than fertilizers containing NH_4^+.

Nitrogenous fertilizer should be applied in split doses to reduce nitrogen losses through volatilization and denitrification. The required quantity of phosphorus and potash along with first dose of nitrogen (one third for pearl millet, wheat and barley and half for cotton and mustard) should be applied at or before sowing. The remaining nitrogen should be applied in two equal splits at first and second irrigation i.e. 25 and 45 days after sowing in pearl millet, wheat and barley and in one dose in cotton (60 days after sowing) and mustard (30 days after sowing). Under limited water supply, nitrogen should be applied in two equal doses at sowing and first irrigation in wheat and barley.

Table 5.13 Effect of soil salinity and nitrogen levels on grain and straw yield of wheat (t/ha) in saline soil

Treatment	1984-85		1985-86	
	Grain	Straw	Grain	Straw
Soil salinity (EC_e dS/m)*				
3.95	4.94	8.56	4.27	5.79
5.63	4.08	7.05	3.94	5.28
7.21	2.50	4.70	3.59	4.65
CD 5%	0.39	0.45	0.34	0.52
Nitrogen (kg/ha)				
80	3.45	5.55	3.56	4.74
120	3.84	6.86	3.88	5.21
160	4.24	7.89	4.35	5.77
CD 5%	0.39	0.45	0.34	0.52

*Mean of 1984-85 and 1985-86 at 0-20 cm soil depth at sowing

Foliar application of N along with soil application is economical and beneficial in saline soils. It also saves one irrigation, which is otherwise needed to dissolve and effectively distribute the top dressed fertilizer.

Several workers have observed that conversion of NO_3^--N into proteins is adversely affected by salinity thereby resulting in inorganic N accumulation in plant parts. Therefore, instead of total N content, protein-N should be the true index of N availability in saline soils. In *kharchia* variety of wheat, protein content is not affected up to EC_e of 16 dS/m. In certain parts of Haryana, Rajasthan, Uttar Pradesh and Madhya Pradesh, saline ground water contains appreciable amounts of NO_3^- to be toxic to the plants. Continuous use of such waters delays maturity and adversely affects grain quality. In such areas, canal water in conjunction with ground water helps to minimize the adverse effects.

Positive effects of phosphatic fertilizers on crops yield in saline soils have been observed even when soil contained high amounts of available P. It is attributed to the fact that soil adsorbs more P at higher than at lower EC. Thus, less P will be available in a saline soil solution. Besides, root growth, the effective root absorbing surface area would be much less resulting in lesser uptake of P, a relatively immobile nutrient in the soil. Uptake of P by plant roots may also be adversely affected by competitive uptake of Cl^- ions as both these anions are absorbed by the same mechanism. Thus, in saline soils, phosphate application would enhance the yield by directly providing the phosphorus and by decreasing the uptake of toxic elements like Cl^- and F^-. The yields of wheat and mustard increased significantly with the application of 30 kg P/ha as compared to no application of phosphorus (Singh *et al.*, 1990; 1992) (Table 5.14). Application of phosphorus along with FYM is still more beneficial as this combination markedly alleviates salt injury symptoms in crops.

Table 5.14 Effect of different levels of soil salinity and phosphorus on grain and straw yield of wheat (t/ha)

Treatment	1986-87		1987-88	
	Grain	Straw	Grain	Straw
Soil salinity (EC_e dS/m)*				
3.18	3.99	5.38	5.18	6.88
4.29	3.67	4.92	4.87	6.47
6.52	3.39	6.47	4.58	6.09
CD 5%	0.33	0.49	0.36	0.51
Phosphorus (P_2O_5 kg/ha)				
0	3.32	4.44	4.51	5.99
30	3.78	5.02	4.93	6.55
60	3.94	5.30	5.19	6.90
CD 5%	0.33	0.49	0.36	0.51

*Mean of 1986-87 and 1987-88 at 0-20 cm soil depth at sowing

Saline soils are generally medium to high in available K but could be low in certain situations where excessive leaching has taken place. At high salinity levels plants generally experience disturbance in Na/K or Ca/K ratios and may show K deficiency caused by antagonistic effect of Na, Ca or Mg. On moderately saline soils, application of potassic fertilizers may increase the crop yield either by directly supplying K or by improving ionic balance through the uptake of Na, Ca and Mg. Under high salinity, it may be difficult to effectively exclude Na from the plant by use of potassic fertilizers and, therefore, effects would generally be on account of increased availability of the K. Application of potassic fertilizers (KCl) has been found to increase crop yields and N use efficiencies in saline soils. Potassium (K^+) and chloride (Cl^-) ions are particularly effective in osmotic adjustment by the leaves as both these ions are taken up very rapidly and can be accumulated in high concentrations in plants. For this reason, chloride salinity is sometimes considered less detrimental than sulphate salinity. Fertilizer requirement of various crops in saline soils is reported in Table 5.15. Micronutrient (Fe, Mn, Cu, Zn) deficiencies have not been widely reported in saline soils.

Table 5.15 Fertilizer requirement (kg/ha) of different crops in saline soils

Crop	N	P_2O_5	K_2O^*
Wheat			
Assured fertilizer supply	120-150	30-60	40-50
Under fertilizer constraints	80-100	30-40	20-25
Late sown irrigated wheat	80-100	30-40	20-25
Barley	100-120	30-40	30-45
Mustard	100-120	30-40	30-40
Cotton	60-80	30-40	30-40
Pearl millet	100-120	30-40	30-40

* Potash should be applied on the basis of soil test values

Organic Manures

The beneficial effects of organic/green manure to supply macro-nutrients and to improve soil structure and permeability are well known. Some additional advantages of incorporating organic manures in saline soils are as follows:

- As stated previously, the N losses through volatilization of NH_3 are relatively more in salt affected soils. Thus, FYM/ green manure can serve as temporary binding agent for the ammonical pool of nitrogen to reduce losses. Experiments show an increased response to N-fertilizers in the presence of organic material which suggests their role in reducing the volatilization losses and enhance N-use efficiency.

- Because of small and less active microflora in saline soils, mineralization of organic nutrient fractions is comparatively less. So the retention of nutrients in organic forms for longer periods will guard against their leaching and other losses.

Organic materials *viz.,* press mud (10-15 t/ha) and farmyard manure (25 t/ha) improve the availability of essential nutrients such as Zn, Fe, Ca and Mn. In calcareous soils, the organic manures moderate the pH, electrical conductivity (EC) and exchangeable sodium percent rendering the soil more suitable to grow sugarcane (Sundara and Vasantha, 2004).

Green manuring is another useful practice in the management of saline soils or saline water irrigated agriculture. Therefore, a green manure crop should find an important place in crop rotation. Growing a green manure crop in reclaimed saline soils during summer is also useful because of the following two reasons:

- It provides a crop cover and therefore, resalinisation of reclaimed soil is avoided. The irrigation water applied to crop also manages to keep the salt down.
- The crop when turned over into the soil, adds nutrients and organic matter which invariably helps in improved crop yields.

Amongst several green manure crops, the efficiency of *Sesbania aculeata* (*dhaincha*) has been observed to be more than sun hemp and *guar*. Gupta (1983) reported that *dhaincha* grows well and provides complete cover to the soil compared to sun hemp which could not grow well in saline soils. It may, however, be noted that application of organic/ green manures as a primary source of nutrients may fail to meet all the nutritional requirements of crops. Hence, only a well mix of both inorganic and organic manures is recommended to maintain higher yields and sustained production under saline conditions.

Irrigation Water Management

The time, frequency and quantity of irrigation water to be applied depends largely on the water requirement of crops, climatic conditions, soil characteristics including texture, infiltration rate, moisture-tension relationship, irrigation methods, water quality and depth to water table. However, depending on the availability of water and weather conditions, the irrigation requirements of different crops may change (Table 5.16). In saline soil, irrigation practices must take care of the osmotic stress caused by the high amounts of salt present in the soil profile. Proper aeration in the root zone may be yet another factor which may modify the irrigation practices in saline environment. Thus, an irrigation method should minimize deep percolation losses to avoid abrupt rise in the ground water table in waterlogged saline areas.

Table 5.16 Irrigation scheduling of different crops based on the availability of water

Crops	Number and growth stages for irrigation			
	1	2	3	4
		Border Irrigation		
Wheat	Between CRI and Tillering (30-35)*	CRI and Booting (70-75)	CRI** Booting Milking (110-105)	CRI tillering (40-45) Booting Milking
Barley	- Do- (30-40)	Tillering Flowering (60-65)	Tillering Late Tillering (40-45) Milking (80-85)	CRI (20-25) Late tillering Flowering Milking
		Furrow Irrigation		
Mustard	Flowering (35-40)	Pre flowering (25-30) and Pod formation (55-60)		
Cotton	Cotton requires four to five irrigations. The first irrigation should be given three to four weeks after sowing and the subsequent irrigations at two to three weeks intervals, depending upon the nature of the soil and weather conditions. Flowering (60-65) and boll form-ation (80-85) are critical stages from irrigation point of view. Inadequate moisture at these stages leads to heavy shedding of flower buds resulting in low yield.			
Pearl millet	Generally one or two irrigations are enough. Pearl millet does not tolerate waterlogged conditions, therefore, water stagnation should not be allowed in the field for more than one day.			

* The figures in parentheses indicate number of days after sowing.
** CRI - Crown root initiation

Since, most crop plants are susceptible to salt injury during germination or early seedling stages, an early-season application of good quality water may provide good enough conditions for the crop to grow through its most injury-prone stage. It would not only fill up the root zone with relatively good quality water but would also leach down some salts from the upper 0-30 cm of soil. Thus, wherever, fresh water is available, pre-sowing irrigation should be given with good quality water. Deep tillage and inversion have been found to help to minimize salinity hazard.

Post sowing irrigation with fresh/saline water also ensures reasonably good germination and seedling establishment due to reduced salt content around the seed and a better moisture regime. However, care must be taken to apply enough water as otherwise salts contained in the surface layer might move right in the root zone. Data from a study conducted at Hisar revealed that a heavy post-sowing irrigation (10 cm) yielded significantly more yield than a light irrigation (5 cm). When a post-sowing irrigation is applied, there are chances that crust may form at the soil surface. This must be avoided as otherwise the benefits of post-sowing irrigation may be nullified.

Most plants require a continuous supply of readily available moisture to grow normally and produce high yields. After irrigation, the soil moisture content is high and the salt concentration or the osmotic pressure of the soil solution is minimal: both favorable for crop growth. As the soil progressively dries out, the moisture content goes on reducing while the concentration of salts in the soil solution goes on increasing. Since it increases both the matric and osmotic stresses, plants experience greater difficulties in absorbing water and nutrients from the soil solution. Thus, infrequent irrigation aggravates salinity effects on growth. On the other hand, with frequent irrigations, soil moisture content remains high and does not allow the build-up of salt concentration in the soil solution which ten to minimize the adverse effects of salts. For these reasons crops grown in saline soils must be irrigated more frequently compared to crops grown under non-saline conditions so that the plants are not subjected to excessively high soil moisture stresses (Fig. 5.1, Fig. 5.6). Salts are most efficiently leached from the soil profile under high frequency irrigation (shorter irrigation intervals).

Selection of a suitable irrigation method is of vital significance. A field irrigation system must ensure the supply of water to the plants at the right time in appropriate quantity and quality and at places where the plants need it for their optimum growth. Surface irrigation method includes free flooding, check basin, border strip, furrows and corrugations. In most surface irrigation systems, one cannot apply less than 7-10 cm of water per application unless precise leveling is done and system is scientifically designed. Thus, on-farm management should include proper land leveling and shaping, efficient design and layout of irrigation methods, scientific scheduling of irrigation under both adequate and deficient water supply conditions, irrigation management under high water table and crop planning for optimum water use. Substantial saving in irrigation water could be effected by proper scheduling of irrigation if water table is shallow and ground water is of good quality. Pandey and Sinha (1972) observed that many crops can be grown without irrigation if water table remains within a specified depth.

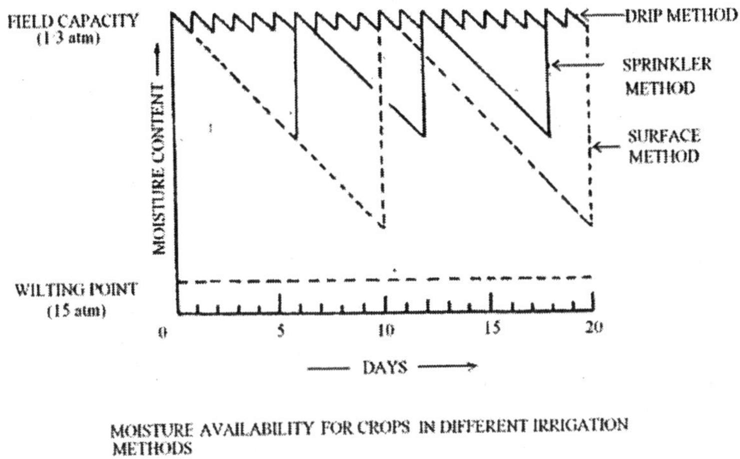

Fig. 5.6 Moisture content variation in irrigation methods as controlled by irrigation frequency

Sprinkler systems, particularly center-pivot and linear-move systems configured with low energy precision application (LEPA) nozzles or properly spaced drop nozzles, could be useful in saline soils particularly to apply post sowing irrigation (Batta, 2011). Drip method, a relatively recent development, could be beneficially applied to saline conditions. Drip irrigation system provides the best salinity management as the salts move to the tip of the wetting front leaving most of the root zone salt free. Subsurface drip irrigation reduces evaporation and pushes the salts to the edge of the wetting front reducing harmful effects on seedlings and plant roots. The method is most suitable for irrigation with saline water as it helps to maintain a zone of low salt concentration and high moisture content resulting in favorable conditions for crop growth.

Land Modifications and Planting Management

When leaching of salts below the root zone is difficult, one should try to live with the salts. It could be achieved by moving the salts away from the primary root zone with irrigation and/or certain crop bedding systems. The goal is to ensure zones of salt accumulation away from the germinating seeds and plant roots. One of the essential pre-requisite of such a kind of management is irrigation uniformity. Without uniform distribution of water, salts might build-up in unintended areas where the germinating seeds and seedling plants might experience reduced growth or possibly complete failure.

The salts in a furrow irrigated field accumulate on the ridges while the area at the bottom of the furrow and around it has the minimum accumulation of salts (Fig. 5.7). Thus, seeds can be planted in the furrows to take advantage of a wet and less saline zone. Moreover, a post-sowing irrigation leaches the salts of furrow soil more efficiently than those of the ridge soil. Dargan and Chhillar (1973) observed that sowing of sugar beet on one side of the ridge half way between the top and the bottom, opposite to the direction of the sun (northern ridge) gave significantly higher yield than flat and ridge sowing. The salt concentration in the root zone under this set up of sowing was comparatively less owing to more accumulation at salts in southern face of the ridge due to direct and intense solar radiation on this face. Similarly, trench method of planting sugarcane gave significantly higher yield than flat sowing. It could be attributed to less concentration of salts, which move to the top of adjoining ridges (Dargan *et al.*, 1973). Seeding on the north-east side of the ridges (Bains and Singh, 1966) proved useful in saline soils, because the salts moved along with water to the drying surface and accumulated there. The EC_e value at south-west, top and the north-east sides of the ridges (60 cm apart, 25 cm high) were 35.4, 24.4 and 6.7 dS/m, respectively. Soil initially saline had EC_e of 15.5 dS/m (Bains and Singh 1966).

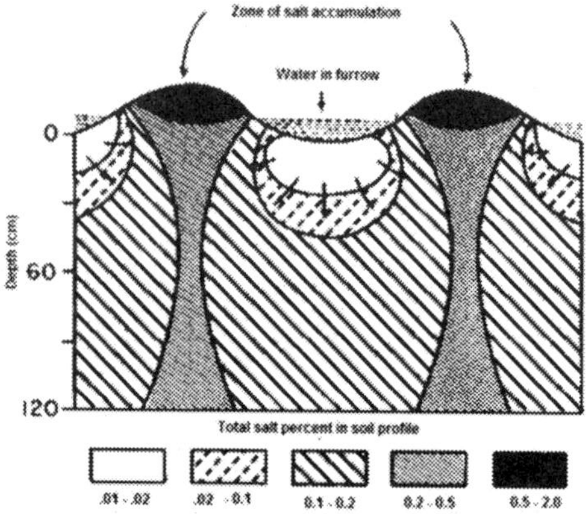

Fig. 5.7 Salt distribution in a furrow irrigation method in saline soils

The method also proved beneficial in saline/alkali soil (EC_e 4.8 dS/m, ESP 55) at Kanpur (Table 5.17). While the data show the superiority of sowing on the side of ridges, even the sowing in furrows gave good results in wheat although some reduction in maize over sowing on the sides of ridges could be observed. It appears that maize being less tolerant to water logging might have suffered due to short-term water stagnation in the furrows. Another difficulty in sowing in the furrows could arise from the fact that the salts may enter the furrows after a rain shower to adversely affect the crop. This situation may not arise if seeds are sown on the sides of the ridges.

Table 5.17 Yield of maize and wheat as affected by method of irrigation and seed placement

Treatment	**Grain yield (t/ha)**	
	Maize (Ganga 5)	**Wheat (Sarbati Sonora)**
Flat sowing	0.93	2.19
Sowing on the side of ridges	1.85	2.51
Sowing in furrows	1.19	2.52
Sowing in 2 m wide beds	0.95	2.35

In a double-row bed system, salt accumulate toward the middle of the bed (Fig 5.8 left). Thus, plants grown away from the middle towards the shoulders of the furrow would encounter relatively low salt stress. Another modification to this method has been in vogue. Here, a notch at the centre of the ridge is made and planting is done halfway between the wet furrows (Fig. 5.8 right). It may be noted that if water is not applied uniformly to furrows (one furrow receiving more or less water than the other), salts might accumulate closer to one side of the bed to affect the plant growth.

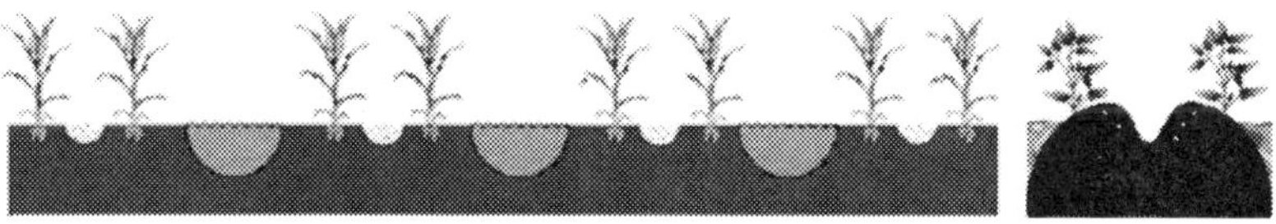

Fig. 5.8 Salt accumulation in double row planting (left) and furrow top with a notch (right) in cases where each furrow is irrigated

In alternate furrow irrigation in single-row bed systems, salts are pushed across the bed from the irrigated side of the furrow to the dry side. Thus, planting towards the shoulder of the wet furrow would be helpful (Fig. 5.9). Care must be taken to ensure that enough water is applied to wet all the way across the bed to prevent build-up in the planted area. This method of salinity management can still result in plant injury if large amounts of rainfall fill the normally dry furrow and push salts back across the bed toward the plants. This phenomenon might also occur if the normally dry furrows are accidentally irrigated. Besides what has been stated, salt leaching of the beds would be necessary after a season or a year as otherwise the beds would accumulate so much salt that it may be difficult to plant them. In slanting ridges, same effect could be obtained by planting the crops on the sides of the ridges as shown in Fig. 5.10.

Fig. 5.9 Salt accumulates on the dry furrow side in alternate furrow irrigation

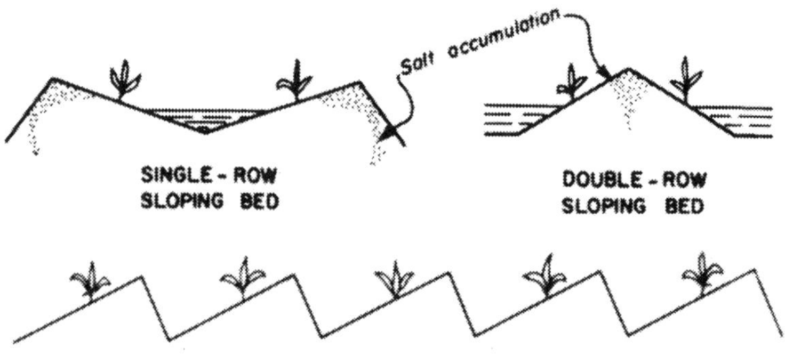

Fig. 5.10 Planting on the sides of slanted ridges to avid salt stress

In drip irrigation salts move to the fringes of the wetting front leaving most of the area under the drip salt free (Fig. 5.11).

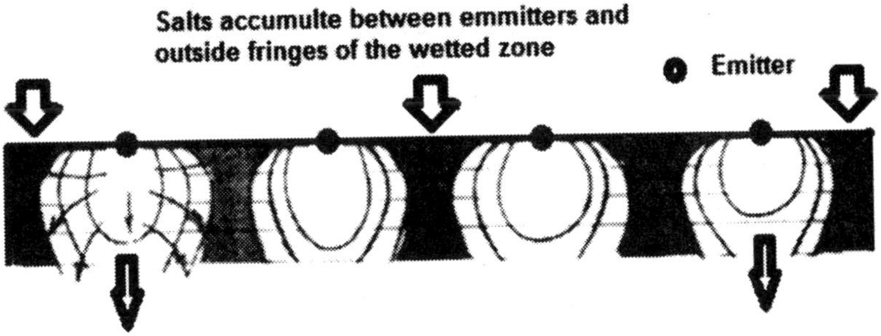

Fig. 5.11 The salt accumulation zone in a drip irrigation system

One of the reasons of variation in crop yields by different irrigation methods is due to variations in salt and water profiles in the crop root zone as modified by the irrigation method. Even the salt and water distribution in space and time during the growth period of crop would influence the yields. It is because of this reason, high frequency irrigations result in low total stress to the plants. Furrow and ridge and flat sowing of turnip under surface and sprinkler methods of irrigation for a saline-alkali silty-loam soil resulted in higher concentration of salts up to 30 cm in the surface irrigation (Yadav and Girdhar, 1977). Although the trend got almost reversed in lower layers, a 25% increase in yield of turnip is obtained in case of sprinkler irrigation, mainly because of lower salinity in the 0-30 cm layer.

Fallowing

Saline water irrigated lands in monsoon climatic conditions are exposed to salinization- desalinization cycles. In arid regions rainfall being inadequate during the winter season, it is unable to meet the water requirement of a crop and therefore, supplemental irrigations with saline ground water are needed to grow a crop. Moreover, desalinization during the monsoon season is also not quite effective because of low rainfall. Under these conditions, it would be appropriate to keep the land fallow during the monsoon season so that salt accumulated during the winter season are leached out making the soil fit to grow the next winter crop. In extremely arid situations, the fallowing may have to be allowed during summer-winter-summer. It means a crop is grown only once in 2 years. Dhir and Bhatia (1975) observed that lands irrigated with waters of less than 5 dS/m EC are generally double-cropped in western Rajasthan, whereas those irrigated with water of higher salinity are used in rotation with a fallow period of varying duration but extended mostly to 18 months in between. It results in natural amelioration of the soil. Salinity changes due to natural rainfall at several locations in western Rajasthan. over a period of 5 years, revealed that the huge salinity build-up during the irrigation cycle is taken care of by one above-average rainfall year or two subnormal rainy seasons (Dhir, 1977). Some salinity is retained at the 40-60 cm soil depth, particularly in clay loam and heavier soils, but this condition does not interfere with reuse of these lands. It was also observed that fallowing was more beneficial than to double-crop the lands.

Mulching

Salt accumulation tends to be greater in bare soils. There is a tendency for the leached salts to return to the soil surface during periods of high evapotranspiration, period between two irrigations and when land is kept fallow. The process is quite fast when the water table is shallow and the salinity of the ground water is high. Thus, any practice that reduces evaporation from the soil surface and/or encourages downward flux of soil water helps to control root zone salinity. Mulching as a means to control evaporation and salinity build-up has been studied in great deal both in India and abroad. Self mulching (soil), stubble or organic mulching and plastic mulching have been used under experimental conditions. In self mulching, land surface is tilled during intervening periods between the harvests of the winter crop to the sowing of the next crop so as to break up the capillaries. The tillage operation by breaking the capillaries, reduces the evaporation and thereby accumulation of salts in the root zone. Additionally, it enhances rain water absorption, thus improving leaching efficiency of the rain water. Crop residues retained at the soil surface reduce evaporative water losses. To be effective fields must have 30 to 50% residue cover. With residue mulching, soils remain wetter and precipitation is more effective to leach salts particularly from the surface soil layer. Damage to crop seedlings being the most due to salinity of this zone, it helps to improve crop productivity. Carter and Fanning (1964) reported significant reduction in root zone salt concentration where cotton-burr mulch was applied at the rate of 90 t/ha. Under straw mulch, a significant reduction in soil salinity was observed resulting in significant increase in wheat yield. Trash mulching was found useful for sugarcane cultivated saline soils as it reduced water losses and thereby minimized the adverse effect of salts present in the soil. When the same trash is incorporated in the soil after mulching, it adds organic matter and nutrients (Sundara and Vasantha, 2004). Plastic mulches in association with drip irrigation effectively reduce salt concentration in and around the drippers. Although, beneficial effects of mulching have been documented, cost effectiveness of the intervention has not yet been worked out. In spite of what is stated before, one can easily say that self mulching and residue mulching could be cost effective mulching techniques to conserve soil moisture and minimize salinity build-up in the root zone.

Weeds, Diseases and Pest Control

The incidences of weeds, diseases and pests in crops grown on saline soils would not be much different than normal soils. As such, weed control and plant protection measures including use of weedicides, fungicides and antibiotics are practically the same as in a normal soil. Timely control and/or taking preventive measures to minimize weeds, diseases and pests problems should ensure better crop productivity of saline soils.

Specific Ion Toxicity

Saline soils besides total salinity as indicated by EC of soil saturation extract might contain several toxic elements. In such cases, crops may suffer more from these toxic ions than the total salinity. Boron, which might be present in most saline soils and water is toxic to many plants at concentrations only slightly in excess of that required for optimum growth. Even small quantities of boron absorbed by the roots are accumulated in the leaves and values above 250 ppm result in typical leaf burns. However, significant reductions in yield of most field crops due to excess boron alone have rarely been reported. Application of gypsum helps to obviate the adverse effects of boron.

Most fruit crops are more sensitive to salinity than are field, forage or vegetable crops. Bernstein (1964) on the other hand pointed out that damage to many fruit crops may be better related to the concentration of specific ions, e.g. chloride, boron or sodium in the soil solution and/or plant leaves rather than to the total soil salinity. Certain stone-fruits, citrus and avocado rootstocks differ in their ability to absorb and transport sodium and chloride ions and have therefore, different salt tolerances. The specific injury symptoms in specific ions sensitive plants appeared before any effect of total salt concentration is observed. For instance, a chloride level of 10 mmol/L in a saturated paste extract is considered toxic to sensitive rootstocks although the same rootstock can tolerate a higher total salinity if it is not due to chloride salts. For this reason, classification of fruit crops with respect to specific ions and according to varieties and rootstocks is important.

ALTERNATE MANAGEMENT TECHNOLOGIES

Amongst many alternate technologies to manage saline soils, growing salt tolerance grasses, afforestation or an appropriate combination of the grasses/plantations is the most appropriate and promising.

Grasses for Saline Soils

Grasses, as a rule exhibit more salt tolerance than arable crops. Selection of a grass for growing on a saline soil depends upon factors such as climate, geographical conditions, the depth to water table and availability and quality of irrigation water. Perennial forage grasses growing close to ground level are ideal from reclamation point of view, because these not only check evaporation but also add organic matter. Bermuda grass is reported to be quite tolerant to salinity (Maas and Hoffman, 1977). Bernstein (1964) reported that yield reduction of Bermuda grass is only 10% at EC_e of 13 dS/m and 50% at 18 dS/m. From field trials conducted at CSSRI, Sampla farm (soil EC_2 8 dS/m, high water table < 2 m in the month of June and ground water EC 22 dS/m), it emerged that Para grass yielded the best when good quality water was used for irrigation (Kumar, 1988). On the other hand, Karnal grass performed the best when saline water was applied to irrigate the grass. On the basis of 4 years average, Gatton panic, Karnal grass and Blue panic also performed well in saline soils. Grasses establish better with rooted slips compared to sowing with seeds when highly saline water is used for irrigation. *Panicum coloratum*, Karnal, Rhodes and Gatton panic grasses were relatively less affected than others owing to irrigation with highly saline water. Yield reduction in *Panicum coloratum* was 50%, Karnal and Rhodes grasses about 65% and in Gatton panic about 75% when irrigated with water of EC 19 dS/m. Hybrid Napier failed to grow with such waters. Anjan and Blue Panic grasses suffered badly even with saline water of 12 dS/m (Kumar, 1988).

Among the nine forage grasses evaluated for their tolerance to irrigation with saline water (EC_i 10 dS/m), *Panicum laevifolium* and *P. maximum* were found to be most suitable species producing 3.43-4.23 t/ha forage annually (Table 5.18). Cultivation of these grasses also ensured that the forage was available during lean period of summer when no other forage crop can be grown. The salinity developed in the soil profile due to irrigations with saline water got leached during the rainy season. Thus, these grasses can successfully be grown with saline water and may form part of silvi-pastoral system for abandoned arid lands.

Ridge and furrow method of planting, 45 cm high ridges spaced at 100 cm between midpoints of two successive ridges, has been found ideal to cultivate *Dichanthium annulatum, Eragrostis, Aeluropus lagopoides* on saline black soils having salinity up to 14-16 dS/m. *Dichanthium Eragrostis* and Leptochloa performed well on ridges and in furrows respectively. Application of nitrogen 46 kg/ha as urea increased the forage yield by about 70% in *Dichanthium annulatum* (Gururaja Rao *et al.*, 1993; 2001). These grasses also responded well to saline water up to 30 dS/m. *Aeluropus* species possessed better forage qualities and had higher salt removal ability.

Table 5.18 Gross dry matter yield (Average of 3 years) and water use efficiency of different grasses

Species	Yield (t/ha) at D_{iw}/CPE			Water use efficiency (kg/ha-cm)		
	0.2	0.4	0.8	0.2	0.4	0.8
Brachiaria mutica	9.54	12.15	11.72	26.7	12.15	11.72
Cenchrus setigerus	4.64	4.57	4.38	12.9	4.57	4.38
Cynodon dactylon	8.91	9.23	10.20	25.0	9.23	10.20
Panicum antidotale	9.34	11.41	11.77	26.2	11.41	11.70
P. coloratum	6.95	10.29	8.93	19.5	10.29	8.93
P. laevifolium	13.49	16.85	16.88	37.8	16.85	16.88
P. maximum (C)	10.87	13.04	12.72	30.5	13.04	12.72
P. maximum (wild)	14.00	14.72	13.72	39.3	14.72	13.72
P. virgatum	9.95	12.10	11.36	27.9	12.10	11.36
Mean	9.74	11.60	11.30	27.3	11.60	11.30

D_{iw}/CPE = Depth of irrigation water/cumulative pan evaporation

Afforestation of Saline Soils

In general, saline lands are more suited to plant fuel wood than timber species. Even the production of fuel wood depends on the ability of the tree species to withstand salinity, drought and various other adverse environmental conditions. Thus, the objectives of the plantations and site conditions determine which plant species to grow. For example *Eucalyptus camaldulensis* can tolerate saline ground water and periodic flooding, but cannot be planted close to the sea coast because of its intolerance to wind blown salt. However, *Casuarina equisetifolia* is able to tolerate the salt spray and provides valuable fuel but is rarely planted inland due to its relatively slow growth rate and poor coppicing ability. Large tsunami affected saline lands along the sea coast have been planted with this specie.

While selecting fuel wood tree specie, the first and foremost criterion is the plant tolerance to the prevailing adverse conditions such as salinity, flooding and poor aeration etc. Wood quality and the potential of the plant to allow silvi cultural practices would also determine the plant's utility. The other functional characteristic such as the plant capability to fix nitrogen, to act as a wind breaker, to provide products for human or animal consumption or other utilities would add to the value of the plant. Ideal specie should not harbor pests and diseases. Species cultivated to lower the water table should have the capacity to intercept and transpire large quantities of water (Burley, 1980).

Much progress has been made in the selection of tree genotypes adapted to waterlogged saline lands. Tolerance of a tree to salinity is governed by the ability of young saplings to withstand salt stress and in areas of shallow water table their ability to withstand excess water stress as well. Fuel wood tree species such as *Prosopis, Tamarix, Casuarina, Atriplex, Eucalyptus, Malaleuca, Acacia nilotica, Acacia tortillis, Salvadora oleoides* and *Parkinsonia aculeata* etc. could be grown. Under moderate barren waterlogged saline lands where salinity is not accompanied by shallow water table, species like *Acacia auriculiformis, Terminalia arjuna* and *Leucaena leucocephala* exhibited good growth but where high salinity and shallow water table co-existed, *Casuarina equisetifolia, Prosopis juliflora* and *Tamarix articulata* grew well withstanding both the stresses. Based on Indian experiences, relative tolerance of some tree species to soil salinity is listed in Table 5.19.

Table 5.19 Salinity tolerance of some fuel wood trees

Tolerant (EC_e 25-35 dS/m)	*Tamarix troupii, T. articulata, Prosopis juliflora, Pithecellobium dulce, Parkinsonia aculeata, Acacia farnesiana*
Moderately tolerant (EC_e 15-25 dS/m)	*Callistemon lanceolatuy, Acacia nilotic, A. pennatula, A. tortilis, Casuarina glauca 13144, C. glauca 13987, C. obesa 27, C. glauca (FRI), C. equisetifolia (FRI), Eucalyptus camaldulensis, Leucaena leucocephala, Erescentia alata*
Moderately sensitive (EC_e 10-15 dS/m)	*Casuarina cunninghamiana (FRI), C. cunninghamiana (Aust), Eucalyptus tereticornis, Acacia auriculiformis, Terminalia arjuna, Pongamia pinnata, Guazuma ulmifolia, Leucaena shannoni, Samanea saman, Albizzia caribea, Senna atomaria*
Sensitive (EC_e 7-10 dS/m)	*Syzygium cumini, S. fruticosum, Tamarindus indicia, Salix spp., Acacia deami, Albizzia guachepela, Alelia herbertsmithi, Caesalpinia eriostachyas, C. velutina, Haematoxylon brasiletto*

Planting Technique and Irrigation Management

For successful tree plantation in waterlogged saline soils using saline ground water for irrigation, there is need to prevent salt build up as well as use of practices to improve aeration in the root zone. Besides, a planting method should help to utilize rain water to the maximum possible extent so that salt concentration in the active root zone of young plants can be controlled to reasonable level. Different specialized site preparation and planting techniques are used to ensure better environment to the young saplings. It has been observed that sowing or planting *Prosopis juliflora* on high ridges is better than planting in pits at little Runn of Kutchh in Gujarat. Field studies carried out by Tomar and Gupta (1985) on seasonal variations in soil salinity and soil moisture due to fluctuating water table indicated that subsurface planting of saplings (below 30 cm depth) could provide less hostile saline environment to the roots. Therefore, they initiated an experiment with ten tree species with two planting methods to assess the relative salt tolerance of tree species (Tomar and Gupta, 1984). In one method, saplings were transplanted in 30 cm diameter auger holes at a depth of 30 cm below the soil surface (subsurface method). In the other, saplings were planted on 40 cm high ridges above the original soil surface. Height, diameter and survival of tree species were invariably more in the subsurface method than the ridge planting (Table 5.20). Earlier studies on salt build-up had shown that the salt accumulation in the 30-90 cm root zone was substantially higher in the ridges than in the natural landscape soil profile. Thus, saplings planted on the ridges remain in disadvantageous position resulting in lower survival and reduced growth.

Table 5.20 Plant height (cm) and survival percentage of tree species at different time

Tree species	Planting methods	Months after planting							
		6		27		69		107	
		H	D	H	D	H	D	H	D
Acacia nilotica	S	95	2.7	326	5.2	586	12.5	641	14.2
		(100)*		69		50		50	
	R	153	3.0	286	4.4	380	6.5	-	-
Prosopis juliflora	S	185	3.1	372	4.5	725	11.7	806	17.8
		(100)		(100)		(100)		(94)	
	R	196	2.9	330	3.7	594	8.4	640	13.5
		(100)		(100)		(100)		(100)	

Acacia tortilis	S	118	2.9	290	4.5	500	10.8	531	10.8
		(81)		(75)		(56)		(50)	
	R	116	1.8	157	2.3	-	-	-	-
		(69)		(38)		(0)		(0)	
Eucalyptus camaldulensis	S	110	2.2	224	3.3	593	8.9	-	-
		(81)		(50)		(38)		(0)	
	R	116	1.8	157	2.3	-	-	-	-
		(69)		(38)		(0)		(0)	
Terminalia arjuna	S	99	2.2	183	4.0	414	12.7	-	-
		(100)		(81)		(50)		(0)	
	R	71	1.1	140	3.2	-	-	-	-
		(100)		(81)		(50)		(0)	
Leucaena leucocephala	S	152	2.4	277	4.7	628	9.2	691	11.7
		(100)		(69)		(63)		(38)	
	R	103	0.7	168	2.0	370	5.4	-	-
		(100)		(19)		(13)		(0)	
Eucalyptus hybrid	S	79	2.2	287	4.1	-	-	-	-
		(63)		(31)		(0)		(0)	
	R	87	1.9	181	2.6	-	-	-	-
		(69)		(31)		(0)		(0)	

S = Subsurface planting method; R = Ridge- trench planting method, H = height and D = diameter; ( ) Figures in parenthesis are percentage of survival of the tree species

Tomar and Gupta (1987) also compared the relative performance of three planting methods i.e. subsurface, ridge-trench and furrow planting. The survival and growth of 5 tree species, 27 months after transplantation revealed that furrow planted trees in general performed better than trees planted with other methods (Table 5.21, Table 5.22). Krepkii (1981) also observed that irrigation by furrows or in circles around the stems (basin) improved the performance of the trees, furrow irrigation being the best. On this basis, it is recommended that furrow planting of trees is most suitable for saline soils.

Cultural Practices

Irrigation is necessary for the survival and growth of plants as adverse effects of salinity get blunted with irrigation. Water application methods that prevent excessive accumulation of salts in the root zone should be preferred. Good quality water for irrigation leaches down the salts from the soil profile. Frequent irrigations with good quality water therefore, are highly beneficial to raise trees in saline soils as irrigation at short intervals provides a low salinity environment to the young roots (Singh *et al.*, 1993). Wherever possible, irrigations should be applied during dry and hot periods. Lack of canal water supplies has often proved a serious bottleneck in the establishment of tree saplings. This has not only hampered the implementation of afforestation programs but has also caused failure of programs aimed at rehabilitating such lands. In the absence of good quality irrigation water, saline ground water can be used in conjunction with fresh water. Even saline water alone could prove to be useful to plant trees. For example trees of the genus *Prosopis* can easily be raised with saline water having EC as high as 20 dS/m.

Table 5.21 Effect of planting methods on the survival (%) of some tree species

Tree species	Survival percentage	
	Subsurface	Furrow
Acacia nilotica	30	100
Acacia auriculiformis	40	85
Casuarina cunninghamiana	60	95
Casuarina equisetifolia	90	95
Casuarina glauca	80	90
Eucalyptus camaldulensis	50	100
Parkinsonia aculeata	75	90
Tamarindus indica	10	50
Terminalia arjuna	15	100

Table 5.22 Effect of planting methods on survival and growth of saplings 27 months after transplanting

Tree species	Plantation methods		
	Subsurface	Ridge	Furrow
Acacia auriculiformis	13 (1.43)	54 (1.48)	65 (2.42)
Acacia nilotica	69 (3.21)	100 (2.85)	95 (2.89)
Casuarina equisetifolia	46 (2.13)	92 (1.11)	95 (3.00)
Eucalyptus camaldulensis	50 (2.24)	44 (1.56)	95 (3.78)
Terminalia arjuna	81 (1.83)	25 (1.36)	90 (2.00)

* Figures in parenthesis express height (m)

Alike arable cropping, high mortality of saplings and limited growth of trees requires that trees are planted at closer spacing than in a normal land (Singh *et al.*, 1993). The exact spacing of course would depend upon the kind of the tree species. It may be noted that a denser canopy resulting from closer row-to-row and plant-to-plant spacing helps to check soil salinization during early growth stage.

Since fertility of waterlogged saline soils is poor and being prone to salt and moisture stresses, even most tolerant tree species do not grow well even when one adopts all the desired management practices. Application of fertilizers to compensate for poor fertility is one way to enhance tree growth. The exact amounts and kinds of nutrients would depend on the kind of salts and salt concentration of the soil. Tomar and Yadav (1980, 1985) reported that application of phosphatic fertilizer significantly benefited the growth of *Eucalyptus hybrid* by counteracting the harmful effect of saline water irrigation. On the other hand, nitrogen application caused a set back to its growth. More studies are needed to critically evaluate the response of fertilizers to growth of tree species in saline soils.

Since, Weeds growth in these soils is limited; weeds should not pose a threat to the planted saplings. However, Weeding operation might help to eliminate the competition for limited moisture. Thus, weeding operation should be carried out as soon as it is felt necessary.

The promising grasses for planting along with trees are: *Aeluropus lagopoides, Brachiaria setigera, Chloris barbata, Echinocloa colonum, Dicanthium annulatum, Leptochloa fusca, Sporobolus helvolus, Cynodon dactylon, Phragmites* spp. and *Sida* spp.

Agroforestry for Saline Areas Near Canals

Salinity problems have arisen in many canal commands because of excessive seepage causing rise in the water table. Trees can act as "biological pumps" as these can transpire huge amounts of water. The companion halophytic shrubs/grasses could be used as ground cover to check the upward flux of salts. Considering soil, agro-climatic and socio-economic factors, trees and companion crops can be chosen from the following.

Trees: *Eucalyptus, Casuarina, Tamarix, Prosopis, Leucaena, Acacia*

Grasses : *Kallar* grass, salt grass, Alkali sacaton, cord grass, Rhodes grass, Para grass

Bushes : Salt bush, blue bush, Kochia

About 200 m wide belt of *Eucalyptus* + Cord grass/*Kallar* grass on both sides of the canal could be grown to intercept quite a large fraction of seepage.

Selecting Tree/Grass Species for Rainfed Saline Lands

About 2/3rd of global land is located in arid and semiarid regions where crop production is dependent upon rains. Practicing agriculture in such areas is always risky. However, if woody perennial plants are integrated with annual agricultural crops, it would reduce the risk of crop failure and would ensure sustainability of biomass production on such lands. Trees of the genus *Prosopis, Acacia, Parkinsonia* are promising for agroforestry for food, forage and fuel wood production. These trees do not compete with the companion crops because of their low water requirement. The yield of grasses and other herbaceous plants also increase when grown in association with *Prosopis* and *Acacia* trees in water scarce situations. The most promising agroforestry systems for arid and semiarid areas are: *Prosopis opuntia, Prosopis cineraria, Cenchrus ciliaris, Prosopis cineraria* - millets, *Acacia albida- millets, Acacia tortilis* - millets. For salt affected soils of the arid and semiarid lands *Prosopis juliflora - Opuntia ficus - indica* (Mill.) is an excellent combination as both have low water requirement and are good biomass producers.

A large area in Saurashtra region of Gujarat is saline and rain fed where only salt tolerant species of trees and grasses could grow. Gururaja Rao *et al.* (2001) identified the following important species which can be introduced into agroforestry systems for such lands.

Trees: *Salvadora* spp., *Prosopis juliflora, Acacia nilotica, Pithecellobium dulce, Parkinsonia aculeata*

Shrubs: *Achyranthes aspera, Boerhavia diffusa, Digera muricata, Kochia prostrata, Suaeda fruticosa, Salsola baryosma*

Grasses: *Aeleuropus lagopoides, Sporobolus helvolus, Cynodon dactylon, Brachiaria rammosa, Dactyloctenium aegyptium, Dichanthium annulatum, Digitaria iliaris*

6

RECLAMATION OF COASTAL SALINE SOILS

The peninsular region of India is bounded by the Arabian Sea on the west, the Bay of Bengal on the east and Indian Ocean to its south (Fig. 6.1). With 8,129 km long coastline endowed with plenty of natural resources, India has one of the most important coastal agro-ecosystem that supports the livelihood security of more than 100 million people. The coastal zone represents the transition from terrestrial to marine influences and vice versa. This zone not only covers shoreline ecosystems, but also the upland watersheds draining into the coastal waters and the near shore sub-littoral ecosystems influenced by land-based activities. The coastal region occupies the deltas, lacustrine fringes, lagoons, coastal marshes and narrow coastal plain or terraces along the creeks. It also includes two distinct major island ecosystems, the Andaman and Nicobar group of Islands in the Bay of Bengal and the Lakshadweep Islands in the Arabian Sea. The coastal agriculture faces several problems, some of them are:

- Climatic hazards (cyclones/floods)
- High humidity
- Saline environment (soil and water)
- Air borne salts
- Shallow saline water table
- Tidal water intrusion
- Heavy rainfall with deep water logging in *kharif* season
- Poor surface and subsurface drainage
- Inadequate good quality irrigation water
- Toxicity of water
- Encroachment of waterways

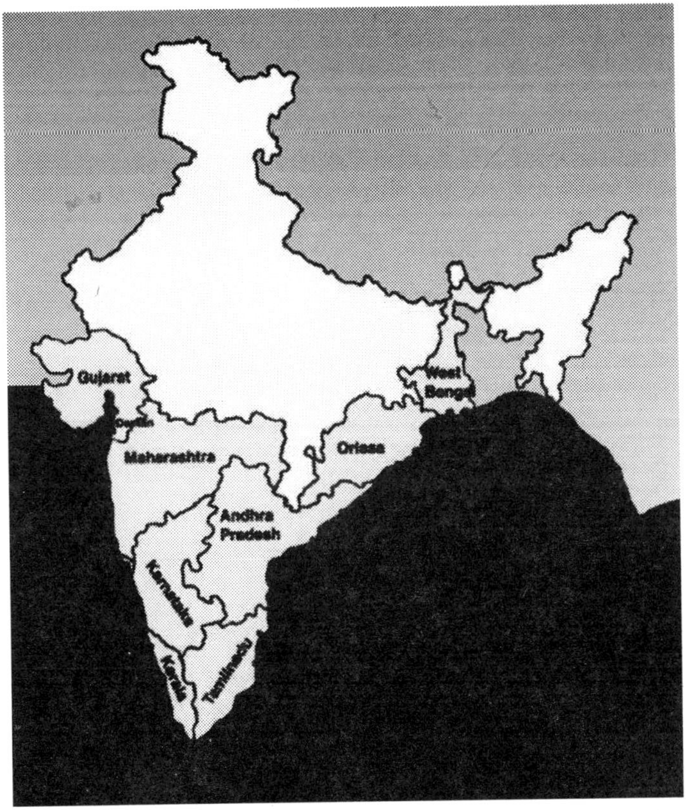

Fig. 6.1 Coastal states and union territories of India

In spite of these hazards, coastal regions support diverse vegetation ranging from rich tropical rain forests to coastal mangroves. The vast deltaic region on the east coast forms the rice-bowl of the country.

GENERAL CHARACTERISTICS

Physiography and Relief

Most coastal areas are stretched between the Eastern Ghats and the Bay of Bengal on eastern coast and between the Western Ghats and the Arabian Sea on western coast (Yadav *et al.*, 1979). Topographically, low lying coastal zone slopes towards the sea and have elevations less than 10 m above the mean sea level (Table 6.1). Most of the coastal zone between the Ganga delta and Cape Comorin is marked by almost plain area whose characteristics have been largely shaped by river deposits. In the deltas of Mahanadi, Godavari, Krishna and Cauvery, the topography is marked by mangrove swamps towards the shore, but further inland, it is the plain country traversed by wide beds of distributaries. The landscape near the western coast indicates the dominance of marine action. The geology from Ganga delta to Mahanadi delta (up to Chilka lake) is largely Pleistocene with pocket of recent alluvium. Almost similar geological features are encountered in Godavari and Krishna deltas. Cambay region in Andaman and Nicobar Islands is marked by low plains.

Table 6.1 Maximum/minimum elevations above mean sea level of some coastal places in India

State	District	Elevation (m) above sea level	
		Maximum	Minimum
West Bengal	24 Parganas	9	2
	Howrah	6	4
Andhra Pradesh	East Godavari	6	SL
	West Godavari	6	SL
	Krishna	3	SL
	Guntur	18	SL
Tamil Nadu	Madras	SL	SL
	Chinglepet	8	SL
	Thanjavur	5	SL
	Ramnathapuram	6	SL
Kerala	Alleppey	19	SL
	Ernakulam	10	SL
	Trichur	10	SL
Maharashtra	----	150	SL
Gujarat	Surat	10	SL
	Valsad	50	SL

SL = Sea level

Climatic Conditions

The climate of the coastal regions does not change much from one place to another. Except for the north Gujarat coast, the area is hot and humid or sub-humid (National Commission on Agriculture, 1976). The general features of the climate are:

- Mild and short winters during December to February
- Hot and humid summers during March to May

- Rainy period during June to September with extended rainy period of October-November in the southern most peninsulas.

Most coastal areas have equable maritime climate throughout the year. The average maximum temperature is in the range of 25 to 35°C. Only West Bengal, Odisha and Gujarat coasts have minimum temperature occasionally reaching below 20°C during three months of December, January and February.

Annual rainfall across almost the entire coastal area excluding the north Gujarat Coast ranges from 820 mm to 3110 mm. Seasonal rainfall and its percentage of the average annual rainfall in different states are presented in Table 6.2. Except south coastal Andhra Pradesh and coastal Tamil Nadu, the areas receive more than 75% of the annual rainfall from southwest monsoon from June to September (Table 6.2). The coastal areas of Gujarat, especially Saurashtra and Kutchh receive much less rainfall. As a result, salinity management is a more serious issue in this region.

Table 6.2 Seasonal distribution of rainfall (mm) at some coastal locations

Region	June-September	October-November
West Bengal	1325 (75)	177 (10)
Odisha	1140 (76)	180 (12)
Coastal Andhra Pradesh	570 (57)	330 (33)
Tamil Nadu	340 (33)	480 (47)
Kerala	2010 (67)	550 (18)
Karnataka	2850 (87)	260 (8)
Konkan	2700 (94)	130 (5)
Gujarat Region	930 (96)	30 (3)
Saurashtra and Kutchh	450 (94)	20 (4)

*Figures in parenthesis are percentage of annual rainfall

Surface wind is generally weak. Strong or high winds are observed mainly in association with depressions and heavy rains which cause severe damage to crops. Cyclonic storms develop in the Bay of Bengal during October-November and strike the east coast causing heavy damage to life and property. In October, the storms largely originate between latitudes 8° to 14° N and move initially in NW direction, but a number of them finally take a curve and move in NE direction. These cyclonic storms together with high tides cause breaches in the protective embankments. Sometimes, over topping of embankments results in inundation of large areas rendering the otherwise normal soils into saline soils

Soil Resources and Salinity

Soils of coastal regions are deep to very deep having coarse sandy to fine loamy texture, non calcareous to calcareous and poor to rich in organic matter. The saline and problematic soils can be divided into three categories; coastal saline soils, acid sulphate soils and coastal sand dunes. In general, the saline soils are slightly to moderately saline while acid sulphate soils are highly acidic in nature. Alkali soils are also encountered in some pockets.

The coastal saline soils occupy an area of about 3.1 M ha in the coastal tracts of West Bengal, Odisha, Andhra Pradesh, Pudducherry, Tamil Nadu, Kerala, Karnataka, Maharashtra, Gujarat, Goa and Andman and Nicobar Islands (Table 6.3; Yadav *et al.,* 1983). This estjmate excludes acid sulphate soils of Kerala which are spread in an area of 0.035 M ha. Any attempt to reclaim coastal saline soils must also address to the general problems described in the previous sections to arrive at cost effective solutions (Nanda and Rai, 1979).

Table 6.3 Extent of coastal saline soils in India (M ha)

State	Area ('000 ha)
West Bengal	820
Gujarat	714
Odisha	400
Andhra Pradesh	276
Tamil Nadu	100
Karnataka	86
Maharashtra	64
Kerala	26
Goa	18
Pudducherry	1
A & N Islands	15
Salt affected area under Mangrove forests	574
Total	**3094 or 3.1 M ha**

The salt affected soils show wide variations, in general being moderate to heavy textured and saline in nature. Alkali soils are found only in a few pockets in Andhra Pradesh and Tamil Nadu. Heavy textured soils (silty clay to clay loam); when dry, crack heavily and look whitish due to salts accumulation at the soil surface. The soils have very low water transmission characteristics, a major limitation in reclaiming these soils. Except some soils of Kerala, majority of the soils have neutral soil pH all throughout the profile. In most cases, the salinity is highest in the surface horizon when sampled in April-May. Analytical data of a soil profile at village Chudamani, in the Balasore district of Odisha showed that dominant anion is Cl followed by SO_4. Sodium is the most dominant cation followed by Mg and Ca. Analysis of the coastal soils of other states also supports that dominant anions are by and large Cl and SO_4 while in most cases CO_3 and HCO_3 salts are absent except in alkali soils. In acid sulphate soils of Kerala, sulphate constitutes the major anion. The soil is acidic to neutral in reaction. A generalized account of the physico-chemical properties of soils at different locations is presented in Table 6.4. Physico-chemical characteristics of a pedon at Upugunduru (Distt Prakasam, Andhra Pradesh) shown in Table 6.5 depict the general characters of a typical coastal saline soil. On the other hand data in Table 6.6 is for an acid sulphate soil at Ambalpuzha (Distt. Alleppey, Kerala). The differences in pH and relative proportions of chlorides and sulphate are noteworthy.

Table 6.4 Physico- chemical properties of coastal soils of India

State	AESR No.*	$pH_{2.5}$	EC_e (dS/m)	OC (%)	CEC {cmol p^+/ kg}	BS (%)
West Bengal	15.1	6.3-7.4	1.5-9.8	0.15-0.52	19.4-22.7	70-80
	18.5	6.5-7.6	4.1-35.0	0.3-0.78	6.6-10.6	74.82
Odisha	18.4	6.3-7.4	2.0-50.0	0.10-0.22	12.8-13.4	75-85
	18.5	4.5-6.3		Tr.-0.20	33.8-55.5	49-69
Andhra Pradesh	18.3	8.0-8.4	4.6-27.0	0.9-1.1	2-6	-
Tamil Nadu	18.1	5.1-6.1	2.0-10.0	0.06-0.09	4.0-6.6	89-100
Kerala	19.3	2.4-3.2	1.0-20.0	9.2-19.9	26.4-58.0	8-57
	19.3	5.1-5.6		0.3-2.6	3.9-7.6	48-58
Karnataka	18.5	3.9-5.7	3.0-10.0	0.07-1.23	0.7-4.6	29-89
Maharashtra	19.3	6.6-7.0	4.0-14.0	0.4-0.7	20.8-28.0	40-45
Goa	19.1	6.3-7.4	4.0-15.0	0.28-0.60	0.5-7.8	48-69
Gujarat	2.2	8.2-8.7	0.20-6.50	0.18-1.10	1.3-3.7	--
	5.3	7.7-8.0	0.46-3.20	0.13-0.15	12.26-19.26	--
Lakshadweep	20.2	8.5-9.1	0.1-0.2	0.11-0.77	0.4-1.5	100

Source: Compiled from several state sources and reproduced from Sen, 2011, AESR, OC, CEC and BS refer to Agro Eco Sub-Region, organic carbon, cation exchange capacity and base saturation, respectively.

General principles of soil salinization have been discussed in Chapter 3. Some of the processes that are most active in coastal areas are:

- Sea water inundation is one of the processes by which soil salinity develops in the coastal regions which may inundate the lands during high tides or cyclones
- Use of saline water traveling to interior parts of coastal plains through rivers/creeks for irrigation
- Use of saline ground water for crop production
- Shallow saline ground water table during the post monsoon season bringing dissolved salts in the root zone through capillary rise more so during the summer season
- Deposition of wind borne salts, the most active process of soil salinization at Kutchh in Gujarat

Table 6.5 Physico-chemical characteristics of a Pedon from Upugunduru (Prakasam) in Andhra Pradesh

Depth (cm)	EC_e (dS/m)	pH_s	Ionic composition (meq/L)							
			Ca	Mg	Na	K	CO_3	HCO_3	Cl	SO_4
0-16	53.1	7.8	91.0	267.0	547.8	1.2	Nil	6.0	848.0	108
16-44	36.1	7.8	47.5	114.0	358.6	1.7	Nil	2.5	476.0	30
44-100	37.2	7.7	38.0	117.5	352.1	1.8	Nil	2.5	502.0	39.9

Depth (cm)	Particle size			Textural class	$CaCO_3$ (%)	OM (%)	CEC (cmol(p^+)/kg)	ESP
	Sand (%)	Silt (%)	Clay (%)					
0-16	43.6	30.4	24.8	1	1.2	1.0	15.3	12.4
16-14	42.6	27.6	28.4	Cl	1.4	0.7	14.1	12.7
44-100	52.2	24.6	22.4	Scl	0.8	0.5	10.9	11.9

l loam, cl clay loam, scl silty clay loam, OM organic matter, CEC cation exchange capacity

Table 6.6 Physico-chemical characteristics of a acid sulphate soil from Ambalpuzha, Kerala

Depth (cm)	EC_e (dS/m)	pH_s	Ionic composition (mmol/L)							
			Ca	Mg	Na	K	CO_3	HCO_3	Cl	SO_4
0-14	2.9	4.1	4.6	8.2	16.2	1.6	Nil	0.5	21.2	7.3
14-32	2.0	4.2	2.5	4.5	12.0	1.5	Nil	0.6	14.5	7.0
32-46	2.1	4.3	3.0	4.5	12.2	1.8	Nil	0.5	13.8	8.5
46-100	1.5	4.5	2.5	3.5	9.5	1.5	Nil	0.4	10.3	6.4

Depth	Particle size			Textural class	$CaCO_3$ (%)	OM (%)	CEC (c mol (p^+) /kg)	ESP
	Sand (%)	Silt (%)	Clay (%)					
0-14	23.9	48.0	28.2	Cl	Nil	3.1	19.2	8.8
14-32	22.9	46.8	29.0	Cl	Nil	1.5	20.8	8.8
32-46	29.7	52.4	19.4	Sil	Nil	0.9	19.6	7.7
46-100	59.7	22.7	17.6	sl	Nil	0.5	10.6	2.4

sl Sandy loam, cl clay loam, sil silt loam , OM organic matter, CEC cation exchange capacity

Similar to inland saline soils, coastal saline soils also follow seasonal salinization-desalinization cycles. A typical seasonal salinity variation in soils at Canning is presented in Fig. 6.2.

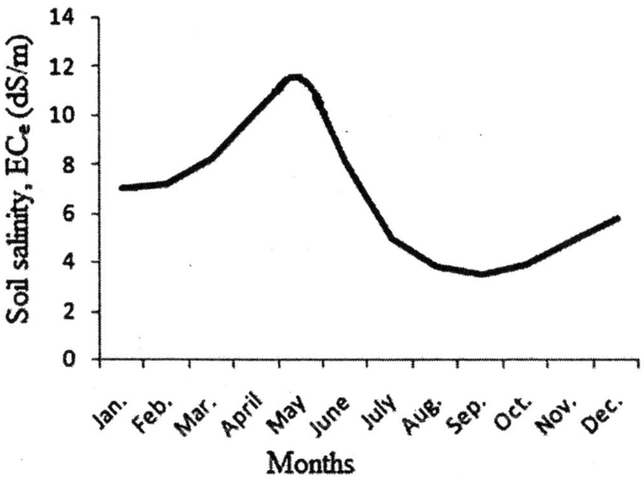

Fig. 6.2 Schematic view of annual variation in soil salinity at Canning

Water Resources

The water resources comprise of rainwater, surface water, ground water and sea water. Surface water resources comprise of sea, rivers, reservoirs, canals, tanks and ponds, lakes and *beels*. Out of the total gross irrigated area, 20-22% is irrigated by tanks. Most areas irrigated by tanks and ponds are in Andhra Pradesh, Tamil Nadu and Karnataka. Restoration of tank irrigation efficiency through desilting, strengthening of the embankments, improvements in field channels and other on-farm water management practices has tremendous potential to augment the food production by 40% in east and west coast. Andhra Pradesh, Gujarat, Karnataka, Odisha and Maharashtra have large area under reservoirs. Most of the lakes, *beels* and derelict water bodies are located in the states of Tamil Nadu and Kerala. The states like West Bengal and Odisha have maximum area under brackish water. Thus, there is ample scope for saline aquaculture in these states. If practices to integrate the use of all the four resources of water are put in place, it will not only increase land and water productivity but also help to manage saline environment of the soil and water.

Ground Water Table Behavior

Depth to ground water table as well as its salinity change with time during the year. Water table is almost at the soil surface in August-September and goes down in May/June depending upon the initial level, rainfall distribution and evaporative demand. Variation in depth to water table recorded at CSSRI, Canning reveal this general trend with water table dropping to about one meter during April/May (Table 6.7). The ground water salinity is minimum during the monsoon season and maximum during the summer months. The dilution of ground water in the monsoon season appears to be responsible for such a variation in the ground water salinity.

The usable ground water potential of the coastal regions is estimated at 21,292 ha m per annum. Only limited efforts have been made so far to utilize this vast resource probably because the sweet water aquifers are very deep and there are geogenic issues which may deteriorate the ground water quality in the long run.

Table 6.7 Depth to water table and salinity of the ground water in different months at Canning

Month	Rainfall (mm)	Depth to water table (m)	Salinity (dS/m)
January	0.0	0.97	10.8
February	11.8	0.95	13.1
March	35.4	0.98	11.2
April	25.2	1.05	10.3
May	369.6	0.53	4.6
June	287.2	0.53	6.1
July	224.6	0.17	4.4
August	224.1	0.00	2.7
September	384.4	0.00	3.9
October	191.4	0.20	4.8
November	76.6	0.56	6.1
December	173.2	0.78	9.0

COASTAL SOILS AND CROP PRODUCTIVITY

The soil salinity in the coastal belt is highly variable reaching a maximum between January to May and decreasing with the onset of monsoon due to leaching, dilution and washing. In addition to salinity, a number of agro-climatic/soil conditions are unfavorable to optimum crop production. Some of the adverse features are as follows (Sen, 2011).

The duration of bright sunshine during the monsoon season is an important factor that adversely affects crop growth. With the onset of monsoon in the month of June and during most of July, the average numbers of sunshine hours are only 3 to 4 per day. It improves to about 7 h in the month of October but is still inadequate from crop production point of view.

The high temperature, a common feature of the coastal regions, affects the growth of thermophyton crops such as wheat, barley, mustard, linseed etc. The high relative humidity in the range of 65 to 96% throughout the year increases the chances of pest's infestation and diseases.

Considerable amount of SO_4 salts present in the soils gets transformed to sulphide producing H_2S gas due to soil submergence during the monsoon season. The initial metal sulphide content of a few mg/L increases to more than ten times after about two months. In saline soils at Canning; 180 to 440 cc of H_2S gas is formed/d/m^2 after two months of submergence. As H_2S at a concentration of 0.07 to 0.1 mg/L in soil solution is toxic to plants, the accumulation of H_2S in soils to toxic levels cannot be ruled out. This may possibly be one of the reasons for poor rice yields in the coastal belt although one may anticipate some reduction in soil salinity due to such transformation of sulphate salts.

TIDES AND EMBANKMENTS

Tides are one of the most important processes affecting the hydrology of coastal zone. The range of tides determines the shore zone and tide height determines the design height of the protective embankments. The tides expose the shore and other parts of the coastal zone to diurnal flooding. The daily tidal range defined as the difference between the maximum and minimum height of water level is about 10 m at Bhavnagar, 4 m at Mumbai, 0.75 m at Cochin and Madras, 1.25 m at Machhlipatnam and 5 m at Canning in West Bengal (Ahmed, 1972). The tidal range is maximum during June-December, adversely affecting land drainage.

Height of Embankments

The coastal saline soils can be cultivated only after constructing protective embankments to protect the lands from flooding due to high tides (Fig. 6.3). The height of the dike is usually calculated on the basis of ground level *vis-a-vis* the water level, height of the high tide expected say once in 10 years or more and the free board. Free board is usually 1 m but may be more depending upon the site condition such as wave rush, wind speed etc. Slope is usually flatter on the sea side usually 3:1 on the sea/river side and 2:1 on the country/ land side. To circumvent the high winds and wave action, shelter belt is usually provided. Wind breaks would be required if problem of shifting coastal sand dunes is anticipated. It may be prudent to provide revetment of the slope when wave rush is expected.

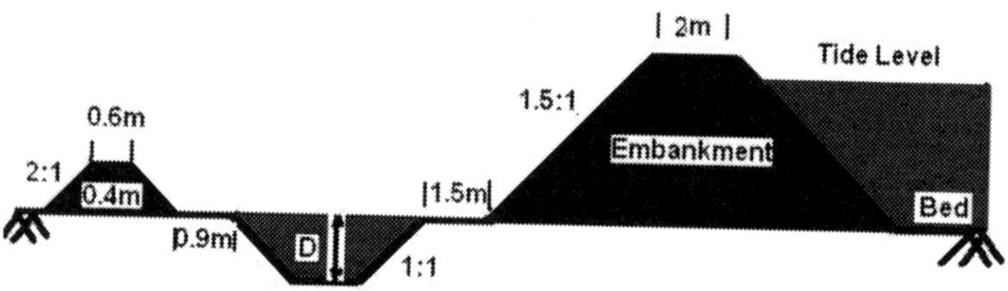

Fig 6.3 A schematic view of an embankment with drain on the land side

Structural Design of Sea Dikes

The design of a dike is site specific and will change with the type of the soils encountered in the area. Besides the elevation, a dike must be structurally strong enough to withstand the forces acting upon it. A structural failure could occur on account of one or more than one of the following factors:

- Direct frontal erosion due to wave attack
- Erosion of landward face following overtopping
- Slipping or slumping of the landward face caused by seepage through the bank
- Building up of water pressure in pervious layers underlying the bank leading to overall instability of the bank

The structural design of the bund should also include the following:

- Slope of the bunds towards the dry and wet side
- Seepage control
- Selection and design of energy dissipaters and revetment for the dike slope on wet side.

It is of no use to construct a high dike with poor structural design or vice versa. A realistic design incorporating the height and structural stability of the dike within the funds available would be a sound proposition

Seepage Drain

Usually the reclamation site is at a higher elevation than the creek and therefore, water level is likely to be deeper in the reclaimed area as measured from the ground surface. When a high tide occurs, the elevation of the water level on the downward side could be shallower in comparison to the ground water table in the area. Thus, it is possible to have a steep rise in the ground water table. The studies made in the coastal areas of Maharashtra and West Bengal (Sahu *et al.*, 1981; Rao and Dhruva Narayana, 1979) revealed that there is no sub-soil water movement during the high tide and thus it can fairly be assumed that no seepage drain along the bund in the upward direction is required in these regions. However, in case such movement is noticed, a suitable seepage drain should be designed and incorporated in the land development plan of the area as is the case in Andaman and Nicobar islands (Fig. 6.3; Gupta *et al.*, 2006).

Embankment Construction in India

Large scale embankment construction works have been undertaken in different states along the coast. The design parameters commonly adopted in Sunderbans are based on the high tide with a 1 m free board. Slopes of 3:1 on riverside and 2:1 on countryside are provided. Top width varies from 1 to 4 m. wherever erosion due to wave action is expected; development of tree belt with mangrove species in the foreshore areas is undertaken.

Sluice Gates

Construction of an embankment to prevent inundation of soils with saline sea water would also prevent the drainage of water unless a drain is provided to remove the excess water generated in the inland areas. This drain however would have two directional flows. While, the sea water through this drain can enter the area at the time of high tide, the water will get out from the land side to the river during low tide. To allow only a unidirectional flow `outfall sluices' are constructed at points where the drain joins a river. The main function of the outfall sluice is to allow free flow of drainage by gravity into the river, estuary or sea during low flood levels and prevent inflow into the drain during high tides. The outfall sluice is similar to an irrigation sluice valve or regulator. The vents, piers, abutments, breast wall, head wall, and aprons are designed for extreme hydraulic conditions as in the case of irrigation sluice. The size and number of vents are fixed in such a way that the drain discharges fully into the river/estuary with a differential head of at least 0.30 m. Proper location and realistic operational schedules help in effective drainage of the inland areas during the monsoon season. The differences amongst various sluice gates are due to the shutters and their closing mechanisms to allow only uni-directional flow to river/estuary. The flap gate type sluice gate is a 90 cm diameter hume pipe with a flap gate opening on one side and can drain about 90 cumecs of water depending upon the available head (Fig. 6.4). Multiple pipe gates are provided at locations where more discharge is expected. In Goa, open sluice gates have been installed until now, but action is being taken to replace them with flap-gate type outlets.

SURFACE DRAINAGE

Most coastal saline soils (except of the irrigated region) face a paradoxical situation. Excess rainwater during the monsoon season causes water stagnation in the crop land due to inadequate drainage. During winter and summer months, there is acute shortage of good quality water for irrigation. From the water balance studies carried out at Canning, West Bengal, it has been observed that during June- October, as much as 450 mm of rainfall is in excess after meeting the losses and demand during that period (Fig. 6.5). The exact quantum of excess water in different coastal states during the same period varies from -300 mm (Tamil Nadu) to 2475 mm (Karnataka). The negative value in Tamil Nadu is mainly due to the bimodal nature of rainfall.

Fig. 6.4 A view of the double purpose flap gate opening on one side

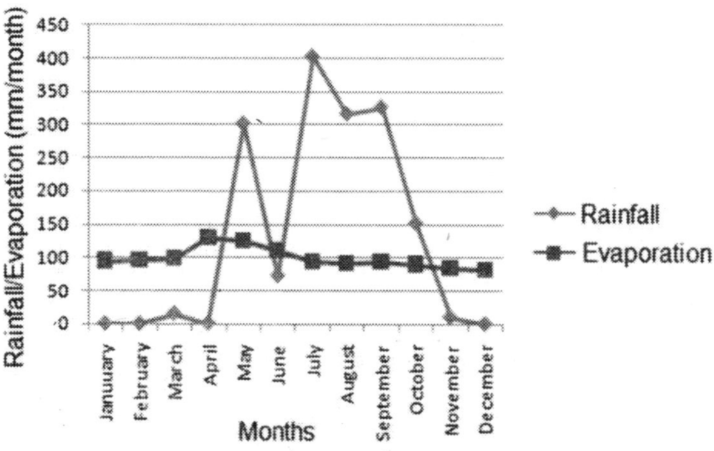

Fig. 6.5 Excess water and its distribution in different months at Canning (2010-11)

Drainage Coefficient

There are several methods to assess surface drainage coefficients for an area. The drainage coefficient recommended by West Bengal Flood Enquiry Commission (WBFEC) is observed to be much lower than the desired value (Rao and Dhruva Narayana, 1979). The Cypress Creek formula, when used for small catchments gives quite high values. For Sundarbans, a drainage coefficient of 37.5 mm/d computed by simplified hydrologic balance appears to be more reasonable (see chapter 2). The drainage coefficient calculated with this approach will remove the excess rainfall of a 5 days storm of 5 years return period within 7 days. The drainage coefficient would reduce to 32 mm/d if losses in the field such as evapotranspiration and percolation, which are about 6 mm/d, are also taken into account. This value seems to be more realistic than the adopted values as recommended by WBFEC. Such site specific studies should be made to work out drainage coefficients for all coastal sites (Ambast and Gupta, 2007).

Drainage of Rice Fields

Most of the area during the monsoon season has standing water depths of more than 20 cm throughout the crop growing period of rice. Sometimes, the depth of standing water exceeds 100 cm. At most places, it is difficult to enter the fields during the monsoon season. To improve this situation, 25-30 cm high peripheral bunds should be constructed to clearly demarcate various fields and the catchment. This intervention helps to control the inflow of excess water from outside the area into the catchment and regulates flow from one zone to the other (Rao and Dhruva Narayana, 1979). It also provides a field level control on the standing water depths maintaining a reasonable (10-15 cm) water level in the rice fields. With channelization of the catchment, excess rainwater from different zones is directly routed to the outlets. In this scenario operational schedules of the open sluice gate during the monsoon season need to be redesigned so that gates open more frequently and for longer periods to maintain the desired water levels in the catchment. It should significantly improve the rice productivity.

Critical Periods for Drainage

The studies on desirable depth of submergence and the rice yields conducted at different places have revealed that the rice plant is susceptible to high depth of water stagnation. The susceptibility of the plant depends on the growth stages at which such water stagnation occurs. The period of transplantation, tillering and panicle formation stages are supposed to be more sensitive to water stagnation. Considering the rainfall patterns, the crop cultivation schedules and the sensitivity of the rice seedling at the transplanting stage, it is concluded that the period between the first week of July to first week of August is most critical period for surface drainage.

Storage and Recycling of Rainwater

Coastal ecosystem in the vicinity of creeks, rivers, deltas, low-lying lands and estuaries is characterized by high salinity of ground water, besides paucity of good quality water in the post-monsoon season. It poses serious impediment in crop cultivation. Although coastal areas receive good rainfall during monsoon, virtually, there is no source of irrigation during *rabi* season. As such, almost the entire area is mono-cropped. The paucity of good quality irrigation water besides build-up of the soil and ground water salinity disallows cultivation of second post-monsoon crop. There is a big potential of storing surplus monsoon rain water for subsequent use during the winter season. The advantages of storing excess rain water are:

- To bring additional area under irrigated agriculture during dry season
- To reduce the depth of flood water in wetland rice fields during the wet season
- To raise the level of the field with excavated soils, if necessary.

Following strategies are adopted to store rainwater to recycle it during *rabi* season for raising the second crop.

- Dugout farm ponds
- Main drainage channels
- Excavated natural channels
- Closed minor rivers and creeks.

Amongst these alternatives, farm ponds appear to be the most promising alternative. Usefulness of farm ponds to provide life saving irrigation to rice crop during September/ October and to grow crops like barley/leafy vegetables/chilli etc. has been established for coastal saline soils of Sunderbans, West Bengal, Konkan region of Maharashtra, and Island ecology of the Andaman and Nicobar. A large number of storage structures have been constructed in the Sunderbans and elsewhere.

Institutional support for this activity is forthcoming because the economic viability of the technology has been proved.

Design specifications for dugout ponds

A farm pond is a multi-purpose structure. While it serves as an outlet for drainage of excess water, water stored in it is used for aquaculture and irrigation. Besides, excavated soil is used to raise the adjoining low lying area of the fields to control runoff from adjoining lands to enter the field. These structures allow the farmers to grow high yielding varieties of rice with better crop management and early sowing of winter/ summer crops (Sen *et al.*, 1996, Sen *et al.*, 2000). Farm ponds are constructed at the lowest portion of the farm to have gravity drainage and to minimize the excavation cost. On the basis of water balance studies at Canning (West Bengal), it has been recommended that a farm pond should occupy 1/5th of the total land holding (Ambast *et al.*, 1998). It means, a farm pond of about 0.2 ha would suffice for a land holding of about 1 ha. It would provide sufficient water for irrigation to crops like chilli, watermelon etc. grown over the rest 0.8 ha. The standard specifications of a farm pond for Sunderbans are:

- Pond size at land surface: 42.3 m x 42.3 m (1,990 sq m)
- Pond depth : 3 m
- Side slope : 1:1
- Depth of water that could be utilized for irrigation: 1.5 m
- Volume of earth to be excavated: 4700 m^3
- Height to which the remaining area can be raised: 50 cm
- Fish culture: bottom 1.5 m of the pond

Minor deviations in these specifications could occur. A schematic diagram of a farm pond constructed in Andaman and Nicobar islands is shown in Fig. 6.6 and the water balance components are reported in Table 6.8.

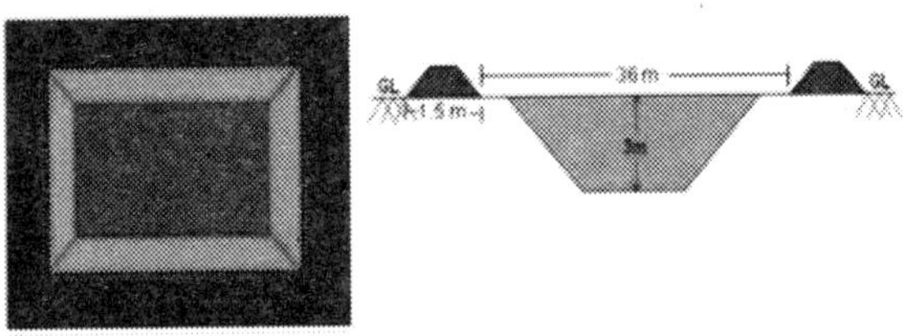

Fig. 6.6 The plan and cross-section of a water pond constructed at Andaman and Nicobar Islands

Table 6.8 Estimated available water in the pond for agriculture

Item	Details
Dimension of the pond	39 m x 39 m
Depth of pond	3 m
Dead storage	0.5 m
Water losses (@5 mm/d for 100 days)	0.5 m
Effective rainfall during summer	0.1 m
Available depth for agriculture	2.1 m
Available storage for agriculture	2500 m^3
Extent of irrigated area	0.85 ha

With this intervention, it should be possible to develop, an improved cropping calendar that allows to keep the land under cultivation throughout the year with regular returns (Fig. 6.7).

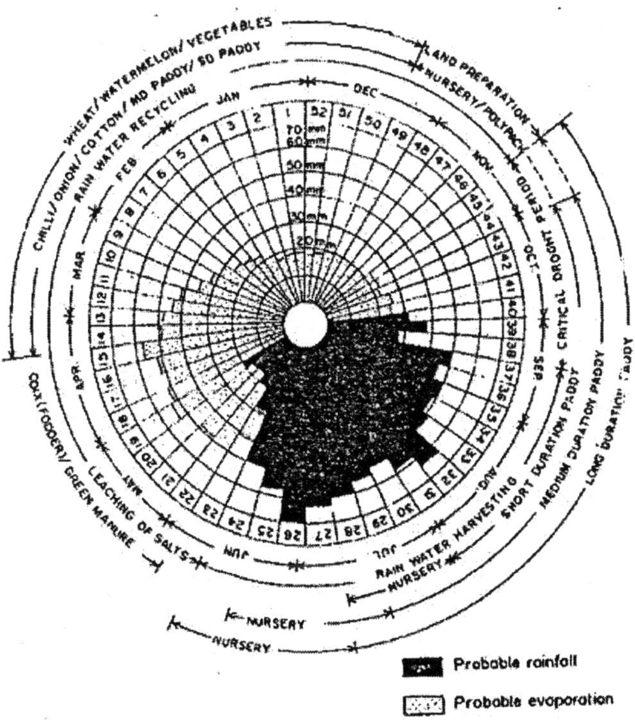

Fig. 6.7 Crop calendar for regular cultivation at Canning

Tapping the creeks

Creeks in coastal areas carry non-saline water during monsoon and post-monsoon seasons but become saline during the late winter/summer. The usual practice to put up a bund in the creek does not help much as water is often contaminated due to breaches in these bunds. Construction of gates with shutters could be a good solution as it will prevent the inflow of saline water into the creek during high tides.

USE OF GROUND WATER FOR CROP CULTIVATION

Although, vast scope exists to exploit ground water for irrigation yet its exploitation is beset with many problems. Therefore, ground water exploitation should always be carried out after thorough hydrological and geological investigations. Cost of such investigations and chances that fresh water would only be available at deeper depths, makes the ground water development a costly affair. Ground water at shallow depths could also be exploited provided its quality is at the most marginal. Considering heavy textured soils and lack of rainfall during winter season, the EC of ground water should not exceed 2.0 dS/m. Another limitation in the rural coastal belts could be due to inadequate and erratic electric supplies. With limited production potential of aquifers, prospects of harnessing cheaper wind energy source could be explored to overcome the cost and constraints of electric supplies.

Dourouv for Skimming Fresh Water

The double cropping in coastal regions is possible only if good quality water is available to meet the shortage of water during winter/ summer months. Since fresh water, being lighter in weight floats on saline water, *Douruvs* in Andhra Pradesh and *Kottais* in Tamil Nadu are used to skim the fresh water floating on the saline water in coastal sands. In the conventional systems, a conical pit is made to skim the fresh water (Fig. 6.8, left). The water is withdrawn from the pit by buckets/pitchers to irrigate the crops. In the improved technology, water is skimmed through a pipeline of about 10-20 m length. The length of the buried drain is designed as per the discharge requirement. Usually two pipelines opposite to each other are constructed. The pit (sump) is constructed to a suitable depth, which is similar to a sump used in subsurface drainage (Fig. 6.8, right). The pipes terminate in the sump delivering the skimmed fresh water. The water from the sump is pumped to irrigate the crops. In the advanced version, sprinkler/drip irrigation is used to conserve the water. Besides the *Dourouv*, multi-point wells have also been used to skim the water. Unlike the *Dourouv*, the mixing of saline and fresh water cannot be discounted in multi-point wells (Raghu Babu *et al.*, 2009).

Fig. 6.8 A schematic view of the improved *Dorouv* technology

RECLAMATION OF SALINE SOILS

Subsurface Drainage

Shallow water table in coastal regions can be managed with subsurface drainage discussed in Chapter 2. The establishment of the water table at some optimum depth would provide proper aeration and minimize secondary soil salinization. The technology has been tested for coastal regions of Andhra Pradesh, Karnataka, Kerala and West Bengal. Open ditches of 1.8 m depth with lateral spacing of 15 m, 30 m, and 45 m were used in a silty clay loam soil at Canning, West Bengal to effectively leach the salts with rain water (Table 6.9). If these drains are properly maintained, water table can be maintained below critical depth, which is about 1.25 m for these soils.

Table 6.9 Salt removal by subsurface drainage at Canning

Spacing of drains (m)	Volume of leaching water collected (m)	Time (days)	Salinity (dS/m)	Salt removed (kg)	Water removed (cm/d)	Salt removed (kg/ha/d)
45	27.86	17.0	7.61	135.67	0.03	13.99
30	31.07	18.5	7.80	155.11	0.05	24.35
15	37.06	19.5	8.33	197.60	0.12	65.80

Encouraging results were obtained at Apikatla, Andhra Pradesh where a subsurface drainage system was installed in farmer's fields covering an area of 7.5 ha. The system installed at an average drain depth of 1 m used stoneware and corrugated PVC pipes with lateral drain spacing of 30 and

60 m. The drain discharge varied with drain spacing being 1.76 to 1.85 mm/d in 60 m spacing and 2.76 to 3.55 mm/d in 30 m spacing during 2007-08. Irrespective of the spacing, the total amount of salts leached during the last 6 years was assessed at 165 tonnes with an average of about 21.5 t/ha. The leaching reduced the EC_e from an initial mean value of 16.2 dS/m to 3.0 dS/m, showing a reduction of about 82% (Fig. 6.9). As such, rice productivity increased from 1.80 t/ha at the time of installation to 6.64 t/ha after six years of installation (Table 6.10), a 270% improvement in yield over control (AICRP, 2009, Ravindra Babu *et al.*, 2010). It seems that a drain spacing of 50-60 m could be used to reclaim salt affected soils. It should also help to reduce the cost of this intervention. Any material i.e. stoneware or corrugated PVC pipes could be used depending upon the availability and cost.

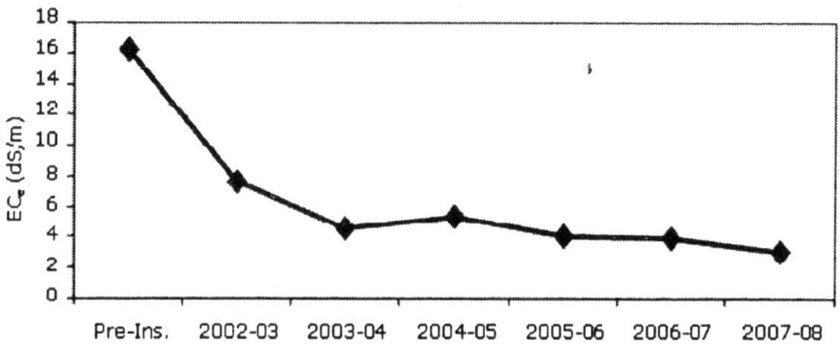

Fig. 6.9 Changes in soil EC_e (0-20 cm) in treated plots

Table 6.10 Paddy yield in subsurface drained land at Apikatla

Year	Grain yield (t/ha)	Straw yield (t/ha)
2002-03	4.10 (128)#	5.50
2003-04	4.24 (136)	5.90
2004-05	6.40 (256)	7.40
2005-06	5.78 (221)	7.32
2006-07	2.09 (16)*	4.94
2007-08	6.64 (269)	7.43

Source: AICRP (2009); # The values in parenthesis are percent increase over initial yield of 1.8 t/ha; * Due to Oogny cyclone at flowering stage of the crop, the paddy yield suffered badly

A major achievement of land reclamation at Apikatla has been to support a second crop in the area. After drainage, fallow area could be planted with a second crop such as sunhemp (*Crotalaria juncea*), pillipesara (*Phaseolus trilobus*), sorghum fodder (*Sorghum vulgare*) and mustard (*Brassica juncea*) with good yields. It resulted in higher cropping intensity and additional net benefits (Table 6.11). Mathew *et al.* (2003) have assembled exhaustive information on subsurfacc drainage for reclamation of acid sulphate soils of Kerala.

Table 6.11 Influence of SSD on yield of different crops following paddy

Year	Fodder sorghum (t/ha)	Pillipesara (t/ha)	Sun hemp (t/ha)	Mustard (t/ha)
2002-03		Germinated but failed		
2003-04	27.20	18.50	19.30	0.27
2004-05	25.40	19.00	20.60	0.28
2005-06	26.10	17.40	20.30	0.29
2006-07	25.20	17.85	18.70	0.67
2007-08	Germinated but failed due to heavy rains during Feb and Mar, 2008			
Mean	26.05	17.85	19.75	0.38

Leaching Practices

Leaching principles and practices are similar to that of inland saline soils. However, some appropriate technologies along with few local practices are briefly described.

Flushing

Unlike inland medium textured soils, heavy textured soils of the coastal region are amenable to flushing. In this system, water up to 10 cm depth is impounded for 7 days. Within 7 days, EC_e of the soil decreases by 4 to 5 times while that of impounded water increases by 5 to 15 times. If the impounded water is drained, it results in substantial reduction in the soil EC. It has also been found that reduction in soil EC is more if plowing/puddling is carried out before draining the impounded water. It is a quite effective method of salt removal, although best results are obtained in conjunction with subsurface drainage. The method is labour intensive and time consuming.

Leaching with continuous ponding

Continuous ponding to leach saline soils is a costly proposition especially in view of limited fresh water in the post monsoon season and shallow water table during the monsoon season. If water is available, continuous ponding in rice fields during winter in the presence of subsurface drainage is an effective way to leach down the salts. In a study, it emerged that soil salinity decreased from 10 dS/m to about 5 dS/m to a depth of 75 cm. Thus, cultivation of winter rice crop with good quality water (EC about 1.5 dS/m) could be a good practice to desalinize saline soils. The only limitation is that sufficient amount of irrigation water should be available during the winter season.

Leaching after addition of amendments

In fine textured soils with poor transmission characteristics, the leaching could be enhanced by addition of soil amendments. Sand, rice husk, rice straw etc. proved effective in improving the physical condition of the soil and thereby resulting in efficient leaching of soluble salts.

Leaching through rainfall

Seasonal changes in the soil salinity have been studied for coastal saline soils at Canning (West Bengal) where average annual rainfall is around 1800 mm. Maximum soil salinity was observed during the month of May while minimum was observed in September. It was observed that nearly 70% of the salts initially present in the soil profile could be leached during the monsoon season. Chavan *et al.* (1981) reported that only 20 to 50% of salts could be pushed below the root zone due to rainfall (Table 6.12). In comparison to arid and semiarid regions, leaching due to rainfall is much less in the humid regions. It could be attributed to high run-off associated with heavy storms, poor rain water management in fields and shallow ground water table.

Leaching and *Pokkali* cultivation

In Kerala, where plenty of rain water is available, leaching is accomplished through flushing of salts. In local parlance, the system is known as *Pokkali* cultivation. The system consists of making mounds 0.5 m high and of about 1 square meter each. The operation is completed by March-April before the onset of the monsoon season. With the onset of the monsoon in May-June, the salts are washed off and the standing water is also drained off the fields. The young seedlings planted on these mounds are saved from total submergence as the monsoon rains lash the area. After sometime, when these mounds are levelled, the clumps of seedlings get fixed due to weight of the soil attached to the clumps.

Although *Pokkali* cultivation is crude and primitive yet it is very peculiar and unique system showing the ingenuity of the people in the past. It involves scientific principles of reclaiming saline soils under the most adverse conditions (Mathew *et al.*, 2003).

Table 6.12 Rainfall and salt leaching in coastal saline soils (Paragaon, Maharashtra)

Annual rainfall (mm)	Salinity (% NaCl)		Reduction in salinity (%)
	May	November	
2984	3.80	2.90	21.05
3248	3.60	2.50	30.55
2342	2.53	1.52	39.92
2898	3.15	1.55	41.27
2879	2.93	1.57	46.42
2026	2.57	2.04	20.62
2674	2.20	1.32	40.00
3368	2.41	1.23	48.96

Crops and Cropping Pattern

In the predominantly mono-cropped coastal tracts, the net sown area in the east coast plains is 8.58 M ha with a cropping intensity of 134%, and in the west coast is 2.77 M ha with a cropping intensity of 125%. Rice is cultivated as rain fed upland ecosystem, shallow lowland ecosystem, intermediate and semi-deep ecosystem and deep water and floating rice ecosystems. The productivity of rice under all the systems is very low. Mono or multiple rice cultivation also impairs soil fertility and productivity. There is a great scope to increase rice productivity in the eastern belt through improved agronomic and other crop management practices. The productivity of shallow water rice having ⅔rd of its share of the total area in the country can be increased from 1-1.5 t/ha to 2.5 t/ha with better management of inputs. Similarly, there is scope to increase productivity of semi-deep water rice from 0.6 to 0.8 t/ha to 1.0-1.2 t/ha through the use of improved varieties. In deep water rice (1.2 m depth), the present productivity level of 0.5 to 0.7 t/ha can be increased up to 1.5 t/ha by manipulating/advancing the time of seeding. Promising water logging and salt tolerant varieties of rice under different situations in the coastal areas are given in Table 6.13.

Table 6.13 Released/identified improved varieties of rice for *kharif* season

Situation	Varieties
Upland situation (0-15 cm water regime) with moderately high soil salinity (EC_e 6.0-8.0 dS/m)	CSR 1, CSR 2, CSR 3, CSR 4, CST 7-1, Canning 7, Jaya, IET 1444, CSR 23, CSR 36
Shallow water (15-30 cm) with moderate soil salinity (EC_e 4.0 to 6.0 dS/m)	Sumati, CSR 6, SR 26B, Dudhesar, NC 1281, NC 678, SMS, Matla, Hamilton, Jhingasail, Gavir Saru, Gopal Bhog
Semi-deep water (30-50 cm) with low soil salinity (EC_e around 4.0 dS/m)	Utpala, CSR 6, SR 26B, NC 1281, Matla, Hamilton, Asfal, Gavir Saru
Acid sulhpate soils	Canning 7, SR26 B, Mashuri
Varieties suitable for flash floods	CSR 4, SR 26B
Rabi/summer season rice	CSR 4, CST 7-1, Canning 7, Ratna

Three salt tolerant rice varieties CSR 23, CSR 36 and BTS 24 were used in the Tsunami affected saline soils of Andaman and Nicobar islands with good results (Fig. 6.10 and Table 6.14). CSR 36 gave the highest yield under both the situations I and II shown in Fig. 6.10 (Ambast *et al.*, 2007). Holding size, generally less than one ha, makes it imperative that the current cropping pattern should be changed. Research and development efforts need to be focused towards rice based multiple cropping systems. This new multiple cropping systems should be compatible with the available resources for sustainable agriculture and livelihood security.

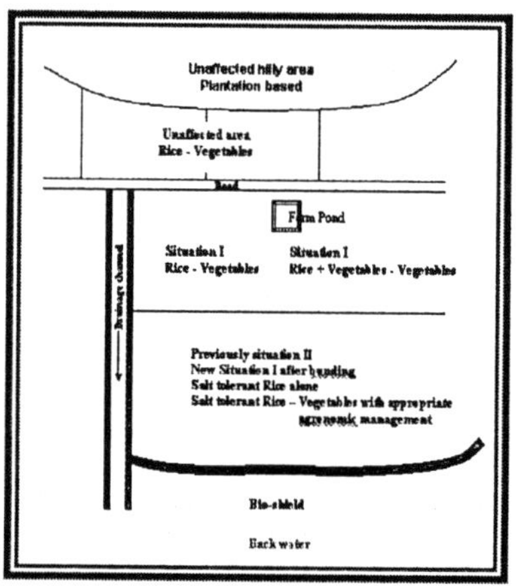

Fig. 6.10 Cultivation of salt tolerant varieties in situations I and II after bunding

Table 6.14 Yield of paddy of different varieties under different situations at model sites

Situation/Location	Varieties			
	CSR 23	CSR 36	BTS 24	Jaya
New Manglutan				
Unaffected land	2.88	3.15	2.57	2.48
Situation I	2.49	2.83	2.24	1.84
Situation II	1.64	1.89	1.26	1.02
Loha Barrack				
Unaffected land	2.83	3.43	-	2.52
Situation I	1.89	2.21	-	1.27
Situation II	1.57	1.76	-	0.95

The International Rice Research Institute (IRRI) in the recent past has identified and transferred submergences tolerance gene (Sub I) in rice varieties which has potential to boost productivity by 2-3 times (Septiningsih, *et al.*, 2009). Similarly, salt tolerance gene (SALTOL) and phosphate (PUP) gene have been identified to improve rice productivity in the coastal belt. These new genetic engineering and biotechnological approaches will play a pivotal role to develop plants tolerant to biotic and abiotic stresses. One such effort is to use the genes drawn from mangroves to develop salt tolerance in arable crops. Dr. M.S. Swaminathan Research Foundation (MSSRF), Chennai has developed a repository of a number of full-length genes isolated from the mangrove species, *Avicennia marina*. Limited field trials of the transgenic rice varieties containing mangrove genes have already begun under controlled conditions (Anonymous, 2007).

Iyengar and Kurien (1971) tested several wheat and barley varieties for cultivation with sea water with or without its conjunctive use with freshwater. The Central Soil Salinity Research Institute, Regional Research Station, Bharuch identified several species of aromatic and medicinal plants, non-edible oil yielding plants and spice crops which can be successfully cultivated in saline Vertisols using saline ground water. Dill (*Anethum graveolens*), a non-conventional seed spice crop gives fairly good yield on saline black soils having salinity of 4-6 dS/m. Salt tolerant fruit/vegetable crops identified for coastal saline soils are:

- Chilli can be grown at EC_e of 6-7 dS/m
- Sapota could be grown up to EC_e 10.7 dS/m whereas guava could be grown up to 5.6 dS/m. Both the crops could be safely irrigated with saline water of 3-4 dS/m

- Salt tolerant tomato (*vars.* Pusa Rohino, Pusa 120, Pusa Early dwarf and Pusa Upahar) and brinjal (*vars.* Pusa Hybrid 6 and Pusa Uphar) can be grown successfully up to EC_e 8.2 dS/m without much loss in yield or fruit characteristics.

A list of salt tolerant crop varieties is given in Table 6.15.

Table 6.15 Improved crop varieties for cultivation in coastal saline soils during *rabi*/summer

Crop	Varieties
Sugar beet	Mariboresistapaly, Rameanskya
Cotton	081, J.K. 260-2, J.K276-4-1-2, F 414, MESR 17
Barley	Karan 19, Ratna, DL 120, K 125, Azad
Chilli	CA 960, Suryamukhi, Pusa Jwala
Safflower	Bhima, Manjira
Linseed	C219-1-1, NP 71
Watermelon	Sugar Sweet, Sugar Baby
Mustard	8-85, T-59, Varun

If salt tolerant products could be exploited for commercial cultivation using seawater, it may open up new and lucrative livelihood options for coastal regions. Exploration and identification of environmental sensitive Genic Male Sterile Lines in tomato, chilli, brinjal and okra should be the top priority of breeders in coastal areas.

Agronomic and Cultural Practices

Following practices are useful in improving rice yield.

- To take care of mortality under the hostile environment, plant population should be increased either by reducing row to row and plant to plant spacing or by increasing number of plants per hill.
- For high yielding semi-dwarf varieties, continuous ponding of water up to 10 cm depth helps to achieve high yield.
- The coastal saline soils in West Bengal are rich in potash and phosphorus. Therefore, there is no response of these fertilizers on crop yields. However, a basal dose of 6-11 kg P/ha is recommended to maintain the soil fertility. Nitrogen @ 80-100 kg N/ha has been found beneficial. Split application of nitrogen in 2-3 splits helps to increase the crop yield. Up to 75% N should

be applied at or before the tillering stage while remaining should be applied at boot leaf stage. Yields of crops and nutrient use efficiency are better with integrated use of organic and inorganic sources of nutrients. Addition of *Sesbania*, *Glyricidia*, city compost, *azola* and farm waste along with inorganic N fertilizer improves productivity and sustainability.

- Slow release nitrogen fertilizers such as sulphur coated urea are useful in improving yield in deep water rice.

Other crop specific practices have also been developed. For barley, 2-3 irrigations of 4 cm each at crown root initiation, tillering and flowering stages are enough to get optimum yield. Chilli requires about 3 L of water per plant in about 4-6 irrigations. Application of 40 kg N/ha in full at sowing time gave highest yield of chilli.

Residue recycling and application of amendments

Spreading rice husk (10 t/ha) on fields and mixing it with surface soil by plowing during the fallow winter periods has been found to improve soil physical condition resulting in greater intake of water and leaching. It also helps to increase the rice yield during the following monsoon season (Table 6.16). Parur (1971) also found that the application of rice straw, green manure and molasses had a beneficial effect on the grain yield of rice.

Table 6.16 Grain yield of rice (t/ha) with rice husk mulching

Treatment	1973	1974	1975
Single rice crop in rainy season (T1)	3.56	4.93	3.02
T1 + additional rice crop during subsequent winter season (T2)	3.59	5.69	4.00
Mulching with rice husk, 10 t/ha during fallow winter period (T3)	4.89	5.29	4.15
C.D. at 5%	0.64	0.36	0.89

Source: Bandyopadhyay and Sen, 1977

In alkali soils of Tamil Nadu, application of amendments such as gypsum and Distillery Spent Wash (DSW) proved effective in reclaiming alkali lands (see Chapter 4). However, application of some source of organic matter with amendments proved to be a better proposition.

Improved Irrigation Practices

Conventionally, surface irrigation methods are more common in coastal areas, although drip and/or pitcher irrigation have several advantages over the surface irrigation methods. These methods suit to saline soils as well as help to utilize saline water for irrigation. Since frequent irrigations are applied with drip irrigation, it blunts the adverse effects of osmotic stress. Moreover, movement of salts to the fringes of the water front leaves the crop root zone less saline. Since these techniques also conserve water, it should be possible to increase winter cropping if improved irrigation technologies are applied on a large scale.

Multiple Cropping in Coastal Saline Regions

A number of multiple cropping models have been developed to utilize every inch of the land and fallow lands during *rabi* season. Intercropping of juvenile phase of oil palm with crops like maize, tobacco, chillies, ridge gourd, okra, banana, colocasia and napier grass yielded encouraging results under West Godavari conditions. Intercropping with banana, spider lily and turmeric was found promising in Gujarat. CTCRI, Thiruvananthapuram (Kerala) showed that under the upland rain fed condition, intercropping of coconut with ginger (variety Varada) gave a net return of Rs.105,940 per ha against Rs.16,990 per ha with coconut alone. At Bhubaneswar, integration of fisheries and plantation crops with traditional monoculture of rice raised the farm income from Rs. 4166 to Rs. 69,275 per ha per annum. Thus, the integrated farming systems have vast potential to improve the livelihood of the farming community with more efficient and eco-friendly resource utilization.

A model based on rice-fish-vegetable cultivation during the *kharif* season has been perfected for the Sunderbans (Fig. 6.11). While fishes utilize the water in channel specially constructed around the field, vegetable crops such as brinjal, lady's finger, bascella, broad beans etc. are grown on the bunds around the field. The model ensures assured regular income to the farmer. To utilize the same area for more returns, highly economical fishes such as prawn (*P. monodon*) and parsia (*L. parsia*) etc. can be cultivated with saline water during winter/summer season. Since the salts accumulated during this period are leached during the monsoon, this model has no adverse effects on the physico-chemical properties of soils.

Fig. 6.11 A view of the rice-fish-vegetable field in Sunderbans

An improved version of multiple cropping is the integrated farming system. It is a land management system in which different components of agriculture like crops, livestock, agro-forestry, horticulture, fisheries, forest trees, mushroom cultivation, bee keeping and sericulture etc. are followed in a holistic manner to increase farmers' income, make best use of available water, soil and climate resources; to generate year round income and employment for the farmers on meager land holdings. System is more remunerative and environmentally safe as it ensures synergic, complementary and efficient recycling of the bi-products like organic wastes/residues. A coconut-based integrated farming system comprising coconut, banana, pepper, mulberry, fodder grass, dairy, poultry and fishery was developed and validated at Central Plantation Crops Research Institute (CPCRI), Kasaragod in Kerala. The output from the system consisted of 23,535 coconuts, 11,445 liters milk, 645 kg broiler birds, 173 kg fish, 17 kg silkworm cocoon, 3,477 quail eggs and 87 kg banana with a net return of Rs.73,142 per annum. In Lakshadweep, coconut-based farming including papaya, poultry, turmeric, ginger, beetroot, radish and onion helped to provide vegetable in the island which otherwise depends on the main land for vegetables. Farm pond based integrated farming model has been proposed for Andaman and Nicobar islands to meet vegetable and fruit requirement within the island (Gupta *et al.*, 2006).

Land Forming Strategies

In a flat topography, lands remain waterlogged with varying depths of water. Manipulation of land shapes to manage water stagnation and avoid zones of salt accumulation has been found useful in coastal areas to pursue a multi-enterprise model for holistic development of lowland agriculture. Original low lying land could be shaped into furrows (3m top*1.5 m bottom*1.5 m depth) and ridges to cultivate paddy–cum-fish in *kharif* and low water requiring crops in *rabi* (Burman *et al.*, 2010). The exact nature of the land shapes also depends upon the original topography and likely water stagnation depth. For example three land shapes comprising of the furrow and ridge to cultivate rice-fish during *kharif* and brackish water fish culture during *rabi*/summer, farm pond based farming with upland, medium land, original land and pond were tried at Canning, West Bengal. The ridges (1.5 m top*1.0 m height* 3.0 m bottom width) and furrows (3m top*1.5 m bottom*1.5 m depth) and shallow furrows (2 m top*1.0 m bottom*0.75 m depth) and ridges (1.0 m top* 0.75 m height* 2.0 m bottom width) were used. It emerged that while benefit:cost ratio was about 1.09 in original land shape with mono-cropping, it increased to 2.33 in farm pond based land modification and was followed by paddy-cum-fish plus field crops during *rabi* with 2.24. In all land shapes B:C ratio was almost double the original land shape model (Burman *et al.*, 2011).

ALTERNATE LAND MANAGEMENT

In some cases, it is neither feasible nor economical to use highly saline soils and water for crop production. Best land use under such situations is to establish plantations for biomass production (NAS, 1990). Preferred tree species for waterlogged saline soils are: *Tamarix articulata, Prosopis juliflora, Acacia nilotica, Acacia tortilis, Feronia limonia, Acacia farnesiana* and *Melia azadirach.* (Tomar *et al.*, 2002). *Casuarina equisetifolia* is the most popular farm forestry tree in coastal India. The tree is salt tolerant, drought resistant, wind resistant and adaptable to moderately poor soils. It could thrive in close proximity to the sea on loose sand, growing within a few meters of high-tide level with its roots in the sea (NABARD, 1998). Besides *Casuarina, Manikara littoralis* (Sea Mahowa), *Terminalia catappa* (Jangli badam), *pongamia pinnata, pandanus tectorius, pandanus learm var. andamanensium* are suitable for erecting bio-shields along seashores (Gupta *et al.*, 2006)

The *Leucaena leucocephala,* has been grown on coastal sandy soil in Pakistan, which survived when 20% seawater was used in the irrigation water, although yields were reduced by 50%. Jojoba *(Simmondsia chinensis)* is relatively salt tolerant, growing near the Dead Sea with brackish water irrigation in Israel. Jojoba plantations are being tested on coastal sand dunes at Zanjmer, Bhavnagar. Quinoa *(Chenopodium quinoa)* germinated in a mixture of 1/3 seawater and 2/3 fresh water, although it posed some problem later at that salinity. *Neem (Azadirachta indica)* seedlings have been grown successfully in Pakistan on sandy soil using irrigation water with approximately 10,000 ppm salt. *Jatropha curcas*, a bio-diesel plant is being tried to green the coastal deserts in Gujarat. Some of the most productive salt tolerant halophytes were shrubby species of *Salicornia* (glasswort), *Suaeda* (sea blite) and *Atriplex* (salt bush).

The degraded lands in arid and semiarid regions are traditionally left for pastures but their forage productivity is low, unstable and un-remunerative. Dagar (2005) observed that the forage grasses like *Panicum laevifolium* (3.43-4.23 t/ha-yr), followed by *P. maximum* (both local wild and cultivated) outperformed the other grasses. Saline irrigation not only improved their productivity by 3-4 times but fodder (about 30%) could also be made available during the scarce periods of April-June. For highly saline black soils underlain by highly saline ground water, use of halophytic grasses like *Aeluropus lagopoides* and *Eragrostis* has been found ideal for bio-saline agriculture. These grasses can be easily grown using saline water having EC in the range of 30-40 dS/m (CSSRI, 2005). *Aeluropus lagopoides* is more acceptable due to its better forage quality. *Dichanthium annulatum* is suitable for saline soils having salinity up to 12 dS/m. These grasses can also be cultivated in association with salt tolerant trees in a unified agro-forestry system. Salt grasses such as *Distichlis,* viny and ground covers such as *Batis* are also highly productive.

Ultra salt affected soils can be used to grow fruit trees like date palm, guava, ber, jamun etc. Coconut is tolerant to salinity as well as to saline water irrigation provided saline water does not stagnate for long periods. Andaman Tall and Katchal Tall can be raised in raised beds with brackish water. Recently, commercial interest has been generated in mangrove palm *(Nypa fruticans)* for alcohol production. This plant grows naturally in Andaman and Nicobar islands (Dagar and Singh, 1994). Dagar and Singh (1994) listed tolerant trees and shrubs, grasses and forbs for saline including coastal saline areas. Mango, passion fruit and pineapple are listed as sensitive while olive, pomegranate and papaya are moderately tolerant (http://www.ussl.ars.usda.gov/pls/caliche/SALTT43). Considering the salt tolerant nature of the plants, besides

Pandanus sp. CARI, Port Blair distributed Alligator apple (*Annona glabra* L.) and *Morinda* sp. to rehabilitate tsunami affected waterlogged saline lands. A compilation of crops that could be irrigated with medium to high saline waters is given in Table 6.17.

Table 6.17 Irrigation water salinity tolerance of some crops

Irrigation water salinity (dS/m)	Kind of crops	Crops
>20 or seawater	Forest and industrial application	*Salvadoa oleoides*, *Terminalia catappa*, *calophyllum inophyllum*, species of *Pandanus*
14	Medicinal	*Lepidium species**, *Withania somnifera* (Ashwagandha)
12	Under exploited species	*Cordia rothii*, *Azadirachta indica*, *Jatropha curcas*, *Adhatoda vasica*, *Ricinus communis* and *Catharanthus roseus*, *Plantgo ovata* (Isabgol)
10	Aromatic and medicinal	Vetiver, Palmarosa and Lemon grass
	Flowers	Calendula (*Calendula officinalis*), Chrysanthemum (*Chrysanthemum indicum*) and German Chamomile (*Matricaria chamomilla*), Periwinkle (*Chatharanthus roseus*)*
8-10	Medicinal	Egyptian henebene, *Cassia angustifolia* (Indian senna)*
	Forestry	*Acacia nilotica*, *A. tortilis*, *A. farnesiana*, *Azadirechta indica*, *Eucalyptus tereticornis*, *Pithecellobium dulce*, *Prosopis juliflora*, *P. cineraria*, *Tamarix articulata*, and *Feronia limonia*
	Grasses	*Panicum laevifolium*, *P. antidotale*, *P. virgatum*, *P. maximum*, *Cenchris ciliaris*, *C. setigerus*, and *Brachiaria mutica*

Compiled from Dagar (2005), * Lodha (2007)

CULTIVATION OF SEAWEEDS AND HALOPHYTES

Seaweed cultivation is picking up fast, being the medical food of the 21st century. Seaweeds grow abundantly and naturally along the Tamil Nadu coast, inter-tidal and sub-tidal regions of Gujarat coasts and around Lakshadweep and Andaman and Nicobar islands. Rich seaweed beds are found around Mumbai, Ratnagiri, Goa, Karwar, Varkala, Vizhinjam and Pulicat in Tamil Nadu and Chilka in Odisha. Amongst all the cultivation methods developed for *G. acerosa*, bottom-culture method using coral stone as a substratum is found to be the best suited.

Recently, large-scale cultivation of *Kappaphycus alvarezii* along the Tamil Nadu coast, has given a crop yield of 25 tonnes (dry weight/ha-yr) for net bag method, 40 tonnes (dry weight/ha-yr) for raft method and 45 tonnes (dry weight/ha-yr) for open culture method in eight harvests. In preliminary experiments at Okha, northwest coast of India, a crop yield of 22 tonnes (dry weight/ha-yr) for the raft method could be obtained (Subba Rao and Mantri, 2006).

Practical applications of the concept of sea farms to bloom the deserts along the seacoast have already begun (Muhawi, 2003). Kutchh area along the sea coast in Gujarat has been identified as a possible site to establish sea farms. Within practical possibilities, the multiple enterprise system to develop sea farms consists of establishing a mangrove forest and cultivation of *salicornia* that provides gourmet vegetable from its young shoots and produces a fine, edible oil and a high protein meal from its seeds. The large amount of biomass produced could be used as animal fodder. Salt lakes are provided for aquaculture. The shrimp species *Penaeus vannamei* has been successfully grown at seawater farms in Eritrea. Technical and economic feasibility of extensive production of artemia, a brine shrimp whose nauplii are used as feed for shrimp and fish larvae, in salt pans without hampering salt production showed good promise. Other potential halophyte is *Salvadora persica.* It can be grown in highly saline soils or irrigation with highly saline water (Table 6.18 and Table 6.19). Seeds of this plant are good source of non-edible seed oil rich in C-12 and C-14 fatty acids having immense applications in soap and detergent industry. It has been estimated that nearly 5,000 ha area has come under *Salvadora* cultivation in Gujarat following the technology developed by CSSRI. Some of the promising halophytes for vegetation of coastal regions are listed in Table 6.20. Like prawn culture, a word of caution would be that environmental issues should not be overlooked in any plan to establish sea farms.

Table 6.18 Seed yield of *Salvadora persica* on saline black soil

Salinity (dS/m)	Seed yield (t/ha)			
	Ist year	2nd year	3rd year	4th year
25-35	1.08	1.27	2.41	2.65
35-45	0.04	1.02	1.35	1.72
45-55	0.60	0.88	1.29	1.58
55-65	0.38	0.74	1.27	1.40
Mean	0.72	0.98	1.58	1.84

Source : Gururaja Rao *et al.* (2003); Oil content varied from 29.3% to 32.8% under different treatments

Table 6.19 Growth as influenced by saline water irrigation and fertilizers in *Salvadora*

Treatment	Height (cm) weeks after treatment		
	2	12	22
T_1 :- No irrigation	5.6	14.6	23.0
T_2 :- Irrigation 10 l/plant at 5 days	7.9	32.9	45.5
T_3 :- Irrigation 20 l/plant at 15 days	9.4	33.2	47.2
T_4 :- T_2 +Ammonium sulphate 100 g	11.6	45.8	64.2
T_5 :- T_3+ Ammonium sulphate 100 g	14.6	47.9	64.8

Source : Gururaja Rao *et al.* (2003), Irrigation water salinity 35-45 dS/m

Table 6.20 Halophytes for costal ecosystems

Plant species/trees	Economic use	Threshold Salt tolerance (dS/m)
Salvadora persica	Seeds oil, forage for camel and goats, medicinal value	45-55
Tamarix erioides	Fuel wood	35-40
Cressa cretica	Fodder	12-15
Suaeda maritima	Fodder, crude protein	35-40
Salicornia species	Seed oil, green, pulp for paper	35-40
Pongamia pinnata	Seed oil, green manure	6-8
Agave americana	Fiber	5-8
Aeluropus lagopoides	Forage	16-20
Arthrocnemum species	Seed oil	35-40

ADDITIONAL PROBLEMS IMPACTING AGRICULTURE

Acid Sulphate Soils

Acid sulphate soils covers a sizeable area in the coastal zone. These soils contain sulfides (mainly pyrites) and become very acidic when sulfides are oxidized to sulphates on drying. The soils usually have a pH below 4. In India, acid sulphate soils are mainly confined in the low lying coastal areas of Kerala and Andaman and Nicobar Islands but have been recently identified in the coastal areas of Sundarbans, West Bengal. Combination of abiotic and microbial oxidation of pyrites causes acidification of these soils. Al and Mn toxicity limits the growth of crops to be cultivated in such soils. These soils are, in general, deficient in phosphorus, Ca and K, whereas the levels of exchangeable Mg may be high. With proper management practices, these soils are capable to support good crops. Some of the strategies to reclaim acid sulphate soils are:

- Liming of the soil @ 3 to 4 t/ha or application of rock phosphate @1.5 to 2.0 t/ha during dry season
- Application of double the recommended dose of P fertilizer (35 kg/ha)
- Green manuring along with lime and P application

Coastal Sand Dunes

As per the 7th approximation classification of soils of Andhra Pradesh, about 2.1 M ha are problematic, of which 0.174 M ha are coastal sandy soils with fresh water lens at shallow depth (Subbaiah *et al.*, 1991). These soils occur in 10 to 15 km wide and 972 km long coastal strip extending from Ichapuram in Srikakulam district to Tada in Nellore district. Since the infiltration of rainfall is quick in these soils, fresh water often floats at a depth of 0.5 to 3.0 m below ground level to a thickness of 3 to 4 m over saline ground water/clay layers. More or less, similar situation exists in 700 km long coastal belt of Tamil Nadu, stretching from Pulicat lake in the north to Cape Comarine in the south. Under this situation, wherever fresh water exists, the farmers tap the fresh water with *Dourouvs* in Andhra Pradesh and '*Oothu Kuzhi*' (Spring) or *Kottai* in Tamil Nadu.

Because of the coarse texture nature of the soils, these areas rarely experience water logging and water congestion. However, sand dunes are deficient in plant nutrients due to extensive leaching. Application of adequate fertilizers is therefore, important to improve the establishment and growth of plants. Application of nitrogen in the form of nitrification inhibitor such as N-serve to minimize nitrogen losses as NO_3-N is recommended. These soils have ample scope for cultivation of multipurpose tree species like *Casuarina equisetifolia.*

Coastal Aquifers and Sea Water Intrusion

The aim of any coastal aquifer management program is to draw maximum fresh water while preventing long-term pollution of the aquifer. Because of the seaward sloping nature of the coastal plains, most aquifers in coastal regions are recharged in inland areas. As such, the fresh water recharge would depress the flow of saline ground water to the aquifers having a direct hydraulic connection with bodies of saline water or with poor saline ground water beneath the tidal waters. However, this natural balance is often disturbed by anthropological interventions and the flow gets reversed. Such a reversal of flow in the aquifer is termed as salt water or sea water intrusion. Some of the factors that could hasten the sea water intrusion are:

- Reversal or reduced ground water gradients due to over exploitation of ground water. Many areas in Gujarat have been ruined as a result of this process.
- Destruction of natural barriers due to construction activities or well drilling.

Coastal areas in Gujarat, West Bengal and Tamil Nadu have suffered salt water intrusion due to excessive ground water pumping (CSSRI, 2004). The depletion in ground water in Pudducherry revealed a drop in coastal areas up to 12 m and inland up to 55 m. Intrusion of seawater into the ground water has been observed up to 4 km inland in the southern parts of the state, 2 km inland in the city and 1 km in the northern side. The seawater intrusion has contaminated local aquifers, leaving them unusable for irrigation and human consumption. Seawater intrusion at few places in Bara tract has resulted in highly saline ground water ranging from 10 dS/m to 100 dS/m. In comparison, the water quality range in Vagra and Jambusar is 1.9 to 39 dS/m and 9.0 to 25.0 dS/m respectively. The ground water salinity in other coastal regions of Gujarat varied from as low as 4 dS/m in rainy season to 45-50 dS/m during summer. The well waters in Andaman & Nicobar Islands showed a salinity of 0.77 to 2.82 dS/m nearly 2 years after tsunami (Gupta *et al.*, 2006). The water quality dynamics studies in Guntur district of Andhra Pradesh revealed that out of eight benchmark sites, quality deteriorated at two and remained more or less same at other sites. Fig. 6.12 illustrates one example for each case of the studied benchmark sites (AICRP, 2007)

It may be noted that lands once subjected to seawater intrusion are difficult to reclaim. However, the situation could be corrected by adopting one or more of the following methods.

- Reduce ground water exploitation
- Increase ground water recharge by artificial means
- Construct physical barriers
- Construct interceptor wells for creating troughs to remove intruding inflow of sea water
- Construct hydraulic ridge by injected water through injection wells

- Construct combination barrier usually combining the trough and the ridge barriers.

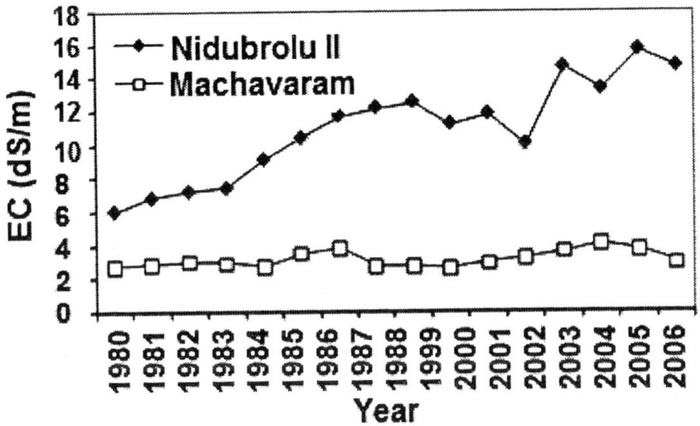

Fig 6.12 Yearly fluctuations in EC of ground water at benchmark sites

Surface and Ground Water Pollution

Ground water in several coastal areas including West Bengal is contaminated with arsenic, which is highly toxic to human and animals. Arsenic contamination of the ground water in West Bengal is traced to the ground water depletion. Several areas are also being polluted through the discharge of industrial effluents loaded with heavy metals. Drinking of arsenic contaminated water and consumption of crops and vegetables grown with arsenic contaminated water is a potential health hazard. Edible parts of leafy vegetable and root crops contain high arsenic compared to grain/fruit crops. Arsenic uptake is reported to increase with P and organic matter application. Its uptake can be reduced by application of 10-15 kg/ha of Fe and Zn.

Some aquifers are contaminated with naturally occurring minerals such as iron, sulfur and fluoride. High levels of fluoride have affected the health of large Indian populace. Direct release of industrial effluents and dumping of solid waste into the sea resulted in exceptionally high concentrations of lead (820 mg/L), mercury (778 mg/L) and cadmium (336 mg/L) in Thane creeks on the Bombay Coast. Sediments along the creeks and offshore stations are also reported to contain significant concentrations of lead. Approximately 5 million tonnes of oil enters the Arabian Sea and the Bay of Bengal receives some 400,000 tonnes each year.

On agricultural front, an assessment made in 1990 revealed that India uses 55,000 metric tons of pesticides a year (herbicides, insecticides and fungicides), of which 25% end up in the sea. An estimated 1,800 tonnes of pesticides enter the Bay of Bengal each year. The increased use of pesticides in some areas has resulted in contamination of shell and finfish. Leaching from landfills and garbage pits transports toxic substances and heavy metals to the water table. Industrial and domestic pollution has severely degraded estuarine and coastal environments. In Gujarat, waste by-products from salt production have killed several hectares of mangrove forests.

Brackish Water Prawn Culture

One of the most important aquaculture practices is shrimp culture which is a traditional practice in the states of West Bengal, Kerala, Goa, Karnataka and Odisha. Shrimp farming received a major fillip with the introduction of the new economic policy of Govt. of India in 1991.While world's cultured shrimp production exceeded 1.6 million metric tonnes in 2003, India is placed fourth with a production of 0.16 million metric tonnes. Major share of the total shrimp production in India comes from Andhra Pradesh (51%), followed by West Bengal (25%) and Odisha (9%). Mud crabs are reported to occupy a second position amongst the cultured species in coastal aquaculture. Species like fin fish have not been yet exploited, though they have tremendous potential to generate foreign exchange. The total potential of coastal aquaculture has been estimated at 0.5 million metric tonnes whereas the total current production is about 0.1 million metric tonnes. Thus, there is ample scope to enhance production and implement species diversification in coastal aquaculture.

Unfortunately, continuous cultivation of prawns in the coastal belt for about a decade now has resulted into serious problems of environmental degradation, deterioration in the soil and water quality and associated socio-economic concerns. Recent studies conducted at Baptala indicated that nearly 3.4 lakh ha area has been abandoned and is lying barren as nothing can be cultivated in these abandoned ponds (Table 6.21). Small and marginal farmers have lost their livelihood as soils have become unfit for cultivation. Adjoining cultivated fields near the ponds being used for prawn culture are affected severely by soil salinization resulting in reduced crop yields. Many farmers after abandoning their ponds have migrated to other areas for employment and to earn livelihood. In many cases, ground water is turning saline. It has been assessed that factory farming of shrimp requires 12 m of saline water/ha and 4-6 tonnes of artificial feed (Shiva, 2003). This salinizes the ground water besides contaminating it as only 16.7% of the feed is converted into shrimp biomass. The rest deteriorates water quality inside the pond and in the ecosystem. The already degraded resources are likely to further degrade because of the following reasons:

Table 6.21 Extent of abandoned aqua ponds in coastal regions of India

State	Abandoned aqua ponds (ha)
Andhra Pradesh	100,000
Orissa	17,200
West Bengal	800
Tamil Nadu	30,000
Puducherry	46
Kerala	210
Karnataka	3,555
Goa	797
Maharashtra	50
Gujarat	188,000
Total	**340,658**

- Increasing exploitation of river water upstream resulting in hydrological alterations causing increased saltwater intrusion. It would convert historically coastal freshwater wetlands and marsh systems to a brackish regime
- Increased exploitation of ground water resulting in seawater intrusion
- Spread of aquaculture activities along the coast leading to increased saline water recharge and pollution of ground water
- Increasing pollution due to diffuse and point sources
- Increased wastewater generation mismanagement disposal related mismanagement.

To restore the livelihood resource of the marginal and poor farmers, the following technology has been developed to reclaim abandoned aqua ponds.

- Leveling of aqua ponds

- Provision of open drainage system for leaching of salts
- Developing source of irrigation water
- Growing *dhaincha* as a green manure crop
- Selection of salinity tolerant paddy variety
- Basal application of $ZnSO_4$ @ 50 kg/ha at time of paddy transplanting
- Nutrient management

With adoption of above package, rice cultivation is possible in lands where nothing grew before. In the experimented fields, rice yields were found to vary from 3.8 to 5.6 t/ha during *kharif*, 2007. The green gram could be grown as the second crop following rice. It may not be out of place to mention that in some cases where soil salinity is too high and leaching water salinity exceeds 3 dS/m, it may take two monsoon seasons to leach out the salts.

Cyclones/Floods

India is exposed to nearly 10% of the world's tropical cyclones. Almost every year on average 5 to 6 cyclones mostly in the Bay of Bengal and east coast are experienced. Two to three of these cyclones could be severe. The records of cyclones for the period 1891 to 1990 showed that while 262 cyclones occurred on the east coast only 33 occurred on the west coast. Most of the cyclones occur during May-June and October-November. Destructive winds are followed by heavy rains causing extensive flooding and soil salinization. Cyclones also result in phenomenal rise in seawater aiding flooding and serious erosion of fore shores as well as pose danger to vessels. The November 1977 cyclone in Andhra Pradesh was the deadliest with 9941 people dead, 900 missing and 34 lakhs rendered homeless, with 1.01 million houses damaged. Crops were completely damaged over an area of 3.3 M ha in a similar cyclone that lashed Gujarat on the west coast in 1998.

The most immediate and visible impact on agriculture is the damage to standing crops, livestock, household property, production assets and physical infrastructure. The extent of damages depends on the intensity of the storm, the area affected, the stage of crops at the time of the storm etc. The extent of damage also depends upon factors such as topography and soil type, moisture holding and drainage capacity, farming practices, permeability of the soil etc. Sometimes, a layer of alluvium gets deposited on fertile lands. It might bury crops and even change the soil quality. In some cases, good top soil may be washed away, exposing less fertile soil layers.

High winds directly impact the forest trees which get uprooted. The forest growth is indirectly impacted due to flood damage, deposition of mud slides and attack by pest and diseases like root and collar-rot favored by waterlogged, oxygen-deficient soil conditions. Nonetheless, forest plantations play a significant role in mitigating damage due to cyclones. Thus, there is a need to establish bio-shield along the cyclone prone coastal areas. Existing mangroves should be properly maintained to serve as wind breaks and provide a buffer zone.

Since cyclones have a vast devastating effect on the populace, there is a need to develop non-structural cost effective methods for rejuvenation of crops/fruit trees during post cyclone phase. For example in the super-cyclone affected areas of Odisha, early rice cv. Vandana cultivated during the pre-cyclone period suffered much less loss than other varieties. Winter potato cv. Kufri Chandramukhi could be cultivated in coastal rice fallow and homestead lands after a cyclone.

Climate Change

The changing climate impacts society and ecosystems in a broad variety of ways. For example climate change may alter rainfall patterns, influence agricultural crop yields, affect human and animal health, cause changes to forests and other ecosystems, or even impact our energy supply. Although climate change would affect the globe as a whole, yet the coastal areas would be affected more than others in a variety of ways. Coasts are sensitive to sea level rise, changes in the frequency and intensity of storms, increase in precipitation, warmer oceans and acidification of ocean waters. For Indian states, a long-term average rise of 1.0 mm/annum in sea level and a projected rise in sea level in the range of 46 -59 cm by the end of twenty first century are forseen (Jena and Mishra, 2011). A recent estimate made by National Initiative on Climate Resilient Agriculture by harmonizing the data reveal that rate of rise could be 1.3 mm/annum (Table 6.22, Singh, 2013). Any change in the sea level would drastically alter the currently established hydraulic equilibrium. While anticipated decrease in rainfall and changing pattern would reduce the fresh water recharge, increasing intensity, duration and frequency of cyclones/other disasters is bound to increase the saline water recharge. In addition, rising atmospheric concentrations of carbon dioxide (CO_2) would result in the oceans to absorb more CO_2 to become acidic. This rising acidity could have significant impacts on coastal and marine ecosystems. Shoreline erosion, coastal flooding, and water pollution would affect man-made infrastructure and coastal ecosystems. While existing challenges are still unresolved, addressing the additional problems caused by climate change may require new approaches to manage land, water, waste, and ecosystems (Sen, 2011).

Table 6.22 Likely variation in climatic parameters on Indian coast

Parameter	**Coastal Region**	
	Eastern Region	**Western Region**
Temperature	Increased by 1.6°C to 2.1°C	Increased by1.7°C to 1.8°C
Rainfall	Increased by 0.2 to 4.4%	Increased by 6 to 8%
Extreme temperature	T Min 2.0°C to 4.5°C	T Min 2.0°C to 4.5°C
	T Max 1°C to 3.5°C	T Max 1.0°C to 3.5°C
Extreme rainfall	No. of Rainy days decrease by 1-5; intensity to increase by 1 to 4 mm/day	No. of Rainy days decrease by 1-5; intensity to increase by 1-2 mm/day
Cyclones	Decreasing frequency / intensity increase	No trend
Strom surge	100 years return period storm surge increased by 15 to 20%	-
Sea level rise	1.3 mm/year	1.3 mm/year

High temperature is known to adversely impact crops yield. Still the cumulative effect of high temperature, increased carbon dioxide and variable rain amounts has yet not been quantified. To manage such a scenario requires altogether different approaches than being currently pursued. Rising sea levels would alter the availability and quality of drinking water in coastal regions by disturbing the interface between fresh and brackish water. To save the human and animal population, instead of building seawalls, it is advisable to initiate a program of protecting mangroves, taking up sheltered plantations along the habitations and preventing destruction of sand dunes, mudflats and lagoons to provide bio-defence against disasters like tsunami and cyclones. Earthen dykes can also be built wherever possible. Promoting Sea Water Farming through agri-aqua farms or promoting below sea level farming as already practiced by farmers in the Kuttanad area of Kerala hold promise to ensure food and nutritional security. Breeding salinity/submergence tolerant crop varieties for cultivation in coastal areas, based on genetic engineering techniques, would play a pivotal role in meeting the challenges of abiotic stresses. Contingency plans need to be kept ready to settle people at safer places. Although it may be difficult to touch all the facets of adaptive strategies, in nutshell, a holistic approach is required to mitigate climate change effects by all agencies and stakeholders through awareness, training, capacity building and adopting more scientific approach towards development.

7

SOCIO-ECONOMIC CRITERIA IN LAND RECLAMATION

Irrigation projects are part of the development programs, which aim at increased and sustainable agricultural production and economic and social change in rural India. On the other hand, many examples could be cited where irrigation projects failed to meet these aims and objectives. Although a number of factors might have caused such failures, a common factor has been the neglect of land drainage in project plans. Inadequate and ill planned drainage resulted in development of water logging, gradual salinization of the soils, and a resultant decline in land productivity. Socio-economic constraints and half hearted institutional support could also be responsible for such a state of affairs. In order that benefits of huge investments made in large irrigation projects are realized, adequate structural interventions such as land drainage must be included in the project plans besides meeting other technical features discussed in the previous chapters. India now faces a stupendous task of reclaiming 6.73 M ha of saline/alkali lands that has already gone out of cultivation and require artificial drainage, chemical amendments and water management. In spite of the fact that techno-economic feasibility of reclamation of such soils has been proved beyond doubt yet the pace of reclamation is not as per expectation. Such a scenario could be attributed to financial, social, administrative and political constraints. In this chapter financial and economic aspects of land reclamation programs are included. It is stressed that all reclamation programs should be governed by sound financial and economic criteria.

FINANCIAL AND ECONOMIC ANALYSIS

Fundamentally financial and economic analysis differs in the following respects:

- What costs and benefits (positive and negative impacts) are included in the analysis?
- How these costs and benefits (impacts) are valued?

The term financial analysis is used to describe a analysis that concerns only with actual monetary flows in terms of cost and returns to specific individuals or groups of individuals within society, i.e. farmers, private firms, public corporations, cooperatives and others. In this sense, financial analysis deals only with those goods and services for which people pay or are paid. It deals with the actual monetary payments e.g. labour, capital and land.

Economic analysis, on the other hand, is concerned with the costs and benefits to society as a whole, regardless of who pays and who gains. It deals with benefits measured in terms of what society is actually willing to pay for goods and services and costs in terms of the opportunity costs involved, e.g. the values of the opportunities foregone when a resource is used for one purpose rather than its next best use that actually would have happened if the reclamation program has not been initiated. It is valid regardless of whether or not actual money exchange hands for a good or service.

The economic analysis, like the financial analysis, is also concerned with profitability, but it is profitability from society's point of view i.e. state region or the nation. In economic analysis, market prices are often adjusted to reflect social or economic values. These prices are referred as accounting or shadow prices. Although many times in practical terms, these are used synonymously yet the differences discussed must be kept in view while interpreting the results.

DISCOUNT RATES AND LIFE

The discount rate (DR) must reflect the cost of the capital to the investor. As such, a discount rate that is commensurate with the interest rate of the banks and/or market rate is taken in financial and economic analysis. Usually the values range from 8 to 12%.

The discount factors (DF) for a given discount rate are calculated by the relation

$$DF = \frac{1}{(1 + DR/100)^{(t)}} \quad (t = 1,2,3...n) \qquad (7.1)$$

Such that t is the year in which cost or benefits accrue and n is the number of years considered in the economic analysis.

Life of a reclamation project is quite long. An alkali land once reclaimed becomes productive forever. For economic analysis of an alkali land reclamation project, costs and benefits for 10 years are taken into account. Similarly, land drainage used in saline land reclamation may have life exceeding 50 years. At Manjri (Maharashtra), a drainage systems installed in 1920 is still operational. However, from an economic or financial point of view, the life of the project is taken as 30 years. The discounted costs and benefits are calculated accordingly as beyond 30 years the discounted costs and benefits do not contribute much to the total.

COST AND BENEFITS

Items of cost in a reclamation project are the following:

- Land acquisition cost
- Depreciation and interest on the investments
- Initial investment in the project including chemical amendments, field collector and main ditches, control works, pipes and pumps, cross drainage works, land development, roads and re-allocation of existing structure
- Investment on replacements in future
- Cost of operation, management and maintenance of the project
- All other associated costs with the project such as outlays for fertilizers, pesticides and implements. Even in the on-going cultivated lands, additional costs may be incurred on account of better varieties of seeds, additional cost of fertilizers, micronutrients and even of the labor. This cost is particularly important when the drainage projects are planned for barren lands or when benefits are measured in the form of increased agricultural production
- Cost of damage to the environment of the area, if anticipated. For example, marine life in creeks could be severely affected with large saline effluent coming out of drainage projects.

In the cost calculations, each item of cost is assigned a value and time. It is an accepted practice to base the calculations on a constant price level. Future inflation is neglected mainly because it is assumed that the value of money declines by that amount. If relative deviations in price or costs are expected, this should be reflected in the calculations.

An accounting procedure is used to convert all prices to a common base by discounting over the economic life of the project by the following relationship:

$$C = \frac{C_t}{(1+i)^{(t)}} \quad (t = 1,2,3....n) \tag{7.2}$$

Here C is the cost at the base level, C_t is the cost in the tth year of the project, and i is the discount rate in fraction. If eq. (7.1) is used, then the factor calculated for each year could be directly multiplied to the total cost likely to be incurred in that year to calculate the discounted costs. The cost components of an alkali/saline land reclamation program are given in Table 7.1 (Gupta *et al.*, 2002).

Table 7.1 Work components as envisaged in reclamation of alkali/saline soil

Sr. No.	Kind of work	Alkali (Quantity/ length/number)	Saline (Quantity/ length/number)
Recurrent cost			
1	Bunding (m/ha)	470	470
2	Irrigation channels (m/ha)	100	100
3	Drainage channels (m/ha)	100	100
4	Outlet (small)	1	1
5	Land leveling (Tractor/h)	10	10
6	Tube well	1 per 4 ha	Canal water, 1 per 4 ha
7	Pump set (no. per tube well or No. for ha)	1 per tube well	1 for 200 ha and/or 1 per ha
8	Soil amendment and application (t/ha)	8-12	-
9	Subsurface drainage (m)	-	200*
Annual costs			
9	Crop production (Rice/pearl millet/sorghum/cotton)	Per ha	Per ha
10	Crop production (Wheat)	Per ha	Per ha
11	Crop production (green manure)	Per ha	Per ha

* Assuming a drain spacing of 50 m. The length will decrease with increasing lateral drain spacing

The relative costs of various components in an alkali land reclamation program reveal that maximum cost is for the application of amendment (Table 7.2). On the other hand in a waterlogged saline land reclamation program, maximum funds are invested on subsurface drainage (Table 7.3).

Table 7.2 Initial reclamation and operational cost of an alkali land reclamation project

Particulars	**Cost (Rs./ha)**	**Percentage of total cost**
On-farm land development	5,000	11.1
Tube well	16,000	35.5
Gypsum application	20,500	45.6
Leaching	3,500	7.8
Total	45,000	100.0

Source: Tripathi and Sharma (2011)

Table 7.3 Drainage and land reclamation costs (1994-95 prices)

Cost component	**Average cost (Rs.)**	**Relative cost (%)**	**Range (Rs.)**
Drainage installation cost	**20,000**	**68.7**	**18,060-22,310***
Pipes	12,200	41.9	12,010-12,930
Envelop material	2,870	9.9	1,660-4,750
Labor	4,930	16.9	4,400-5,980
Connecting costs	**3,250**	**11.1**	**3,250**
Sump	700	2.4	700
Pump	800	2.7	800
Pump house	500	1.7	500
Connecting drain	1,250	4.3	1,250
Land development	**3,370**	**11.6**	**1,250-5,890**
Leveling and bunding	3,370	11.6	1,250-5,890
Other costs	**2,500**	**8.6**	**2,500**
Overhead	2,500	8.6	2,500
Total investment costs	29,120	100.0	25,820-33,870
Annual operation and maintenance	950	-	800-1,150

*Source:*IDNP (2002); *Cost may vary depending upon the kind of materials used

The project benefits accrue due to increased production or due to damage prevented. The benefits of a drainage project may occur on account of the following.

- Increased productivity and production due to reclamation and increase in cropping intensity
- Reduction in cost of farm operations and inputs
- Benefits obtained from timely operations
- Advantages resulting from the introduction of new cropping patterns
- Indirect effect such as on soil structure and hydrology
- Reduced floods, soil erosion and reduced damage due to siltation in and outside the area
- Benefits to public and private properties due to general lowering of ground water table
- Source of irrigation
- Indirect benefits such as employment generation and health etc.

Benefits accrue from increased productivity, increased cropping intensity and crop diversification to a more remunerative cropping pattern is illustrated through the example of a drainage project at Ismaliya. The average yield increased from 23% to 130% over pre-drainage yields over a 3 years period (Table 7.4). Besides, annual cropping intensity in the drained area increased from 73% from pre-drainage level to 150%. Thus, both contributed to increased production over pre-project levels. At a number of places, farmers switched over to wheat where barley was being grown before land drainage. Wheat currently has better marketability and price.

Table 7.4 Percent increase in yield over pre-project levels at two projects sites

Crop	Ujhana	Ismaliya
Wheat	31	89
Barley	24	23
Mustard	--	44
Sugarcane	--	130

Source : Datta and De Jong (1991)

It is very important to include all benefits that accrue even in areas outside the project. For example, ground water table is lowered even outside the actual project area due to increased ground water inflow from the adjoining undrained area. It would result in improved drainage even where drainage has not been implemented. A major difficulty with secondary costs and/or benefits is in assigning a number to these benefits and costs. As such these are not generally included in the financial analysis.

Similar to the costs, benefits are also assigned numbers. The discounted benefits are obtained by the relation:

$$B = \frac{b_t}{(1+i)^{(t)}} \qquad (t = 1, 2, 3 \ldots n) \qquad (7.3)$$

Here, b_t is the benefit obtained in the tth year and B is the total benefits at the same base level as the costs.

The benefits of land reclamation program are calculated in one of the following two ways:

- Pre and post-project comparison
- Comparison of reclaimed and un-reclaimed areas

Both the approaches have strengths and weaknesses but have been successfully utilized. Questions are still there as to which of the two approaches is better. If a project begins with a barren land, comparisons are easier. However, if the land is under cultivation, then the pre project conditions or a control un-reclaimed area should be selected to truly represent the situation. The comparison with the pre-project levels might not yield the true benefits of the project because the spread of water logging and salinity in due course of time will cause more and more degradation of the land and water resources. This deterioration will not be reflected in such calculations. Thus, the true benefits can only be assessed by the second approach. The second approach is also justified because in some cases improvements that occur naturally or otherwise (rather than deterioration) will be reflected in the comparison. The ultimate aim of a comparison should be to bring out the true facts with neither too little nor too much being assigned to the project.

It may be mentioned that as long as the true representative figures are used both the approaches should lead to useful conclusions. A comparison of drained and un-drained and pre and post drainage scenarios for Sampla and Gohana are given in Table 7.5 and Table 7.6. In the former case land being barren initially, remained barren while reclaimed land on an average produced more than 4.2 t/ha of wheat. In the other case, post project productivity was higher than the pre-project productivity. It may be noted that increase in productivity varied from one crop to another, being minimum for wheat. A general decline in productivity in all the crops is seen in undrained area revealing some deterioration in soil quality.

Table 7.5 Changes in soil salinity and crop yield at Sampla (Haryana)

Particulars and Years	Reclamation and yield in different drain spacing (m)			
	25	50	75	Un-drained/pre project
Soil EC_e (0-15 cm) (dS/m)				
0 year	52.2	59.4	49.5	55.2
14 years after	2.1	2.6	4.4	66.0*
Crop yield (wheat (t/ha)				
0 year	0.0	0.0	0.0	0.0
12 years average	4.9	4.6	4.2	**

* After 8 Years, Monitoring stopped, ** Not cultivated

Table 7.6 Impact of subsurface drainage on crop yields (t/ha)

Crop	Area	Before drainage 1995/1996	After drainage 1999/2000	Percent yield increase (+)/ decrease (-)
Rice	Drained	1.41 ± 0.30	1.74 ± 0.24	+ 21.27
(unhusked rice)	Undrained	1.38 ± 0.26	1.28 ± 0.29	- 07.25
Pearl millet	Drained	1.05 ± 0.30	1.34 ± 0.22	+ 27.62
	Undrained	0.92 ± 0.25	0.85 ± 0.28	- 07.61
Sorghum	Drained	23.63 ± 5.44	28.90 ± 4.62	+ 22.30
(Fodder)	Undrained	22.38 ± 4.30	20.68 ± 3.22	- 07.60
Wheat	Drained	03.04 ± 0.61	03.61 ± 0.41	+ 18.75
	Undrained	02.93 ± 0.54	02.52 ± 0.62	- 13.99

Source: Sharma and Gupta (2006)

MEASURES FOR FINANCIAL ASSESSMENT

The financial evaluation of land reclamation projects is accomplished by evaluating few financial parameters. From farmer's point of view, a farmer must get short-term revenue enough to pay his seasonal installment to the bank. It is especially true for bankable projects. Besides, simple rate of return, Benefit-Cost Ratio (BCR), net present worth (NPW) or net present value (NPV), Internal Rate of Return (IRR) and Pay Back Period (PBP) are used in financial analysis.

Simple Rate of Return

The Simple Rate of Return (SRR) of a project is compared with the pre-determined cut-off rate or the minimum rate of return. The following formula is used to calculate the SRR of a project.

$$r = \frac{E - D}{I} \tag{7.4}$$

Here r is simple rate of return on the investment made for the project, E is the annual net benefits expected from the project, D is the annual average depreciation on the fixed assets, and I is the initial capital investment on the project.

Equated Seasonal/ Annual Installment

Cost of the borrowed capital is readily estimated if the interest rate on the loan and the period with in which the loan has to be repaid are known. The equated seasonal/annual installment (ESI) is calculated by the following formula.

$$ESI = (L * I) * [(1+I)^M / \{(1+I)^M - 1\}] \tag{7.5}$$

Here, M is the loan period in seasons (Year*2), I is the interest rate (Interest rate per Annum /2) /100, L is the borrowed capital in Rs.

If the capital is invested from self savings, the cost could be higher depending upon the alternative investment opportunities and should be assessed from the likely returns from the alternative use of the capital. Sometimes, an average interest rate of the borrowed and estimated interest rate of self funds is taken as the interest rate on the total capital for calculating the virtual ESI. For a financially viable project, the seasonal/annual benefits must exceed the seasonal/annual installment. For example; for a soft loan of Rs. 70,000 @ 6% per annum, ESI for 10 years period would be Rs. 8206/-. Thus, the farmers must earn this much amount per season to repay back his loan installment.

Other Financial Indices

Economic criteria come into play as more than one technically feasible project and technical designs could be implemented from the available funds. On the face of it, each appears to be superior to the other. Therefore, to select the most viable project, an economic criterion is applied to all the possible projects and technical solutions. There is no hard and fast rule as to which one of the criteria should be applied. For a project to be viable, benefit must exceed the cost, or the benefit cost ratio should be more than one. Therefore, economic evaluation usually calls for a comparison of the costs and benefits of all the possible projects or alternative solutions within a project.

The financial parameters are calculated by the following relations:

B:C ratio = Total discounted benefits/total discounted cost (7.6)

NPW= Total discounted benefits - total discounted cost (7.7)

IRR= The discount rate at which NPW is zero. While it is easier to calculate the IRR using IRR function in the Microsoft EXCEL, yet it could also be calculated by trial and error using various discount rates. Interpolation could also be made once the values at two discount rates having positive NPW and negative NPW have been calculated.

$$IRR = LDR + DDR \text{ X } \frac{NPW_{LDR}}{NPW_{DDR}} \tag{7.8}$$

Here LDR is the lower discount rate, DDR is the difference between the two discount rates, NPW_{LDR} net present worth at lower discount rate and NPW_{DDR} is the difference between net present worth at the two discount rates.

A check should always be made with the interpolated value. Moreover, interpolation should not be made if the spread of the discount rates is wide.

The payback period (PBP) is given as the number of years it would take for the net undiscounted benefits to repay the investment. Shorter the payback period, lesser is the gestation period and higher the profitability of the project.

Payback period= (Undiscounted initial cost – Σ undiscounted net benefits in each year including the first year) such that the total sum is zero. The year in which the sum is zero is the payback period. If equal benefits are received every year then the payback period is given as:

Payback period = total investments/annual benefits (7.9)

Example Calculations

The calculation procedure to assess financial parameters such as benefit: cost ratio, net present worth, internal rate of return and payback period is given in Table 7.7 using a discount rate of 10%. Discount factors for each year were calculated using eq. (7.1) and are given in column 6. The undiscounted total cost and benefits are given in columns 4 and 8 while discounted costs and benefits are given in columns 7 and 9 (Table 7.7). While benefit:cost ratio and net present worth are calculated using discounted costs and benefits, internal rate of return has been calculated using the function of the EXCEL program under the financial functions. NPV could also be calculated with the help of financial function provided in the Microsoft EXCEL program. Thus, function NPV could also be used to calculate the discounted costs and discounted benefits to determine B:C ratio.

For the example shown in Table 7.7, B:C ratio is 1.77, NPV is 50,760, IRR is 26% and payback period is 5 years. Data in Table 7.8 reveal that payback period for an alkali land program is three years.

Table 7.7 Calculations of B:C ratio. NPV, IRR and payback period for waterlogged saline land reclamation

Year	**Cost**	**O&M cost**	**Total cost**	**DR**	**Discount factor**	**DC**	**Net benefits**	**DB**	**Cash flow**
1	2	3	4=2+3	5	6 (Eq. *.1)	7=4*6	8	9=8*6	10=8-4
1	50000	2250	50000	10	0.909091	45454.55	11000	10000	**-39000**
2	0	2250	2250	10	0.826446	1859.50	12000	9917.36	**9750**
3	0	2250	2250	10	0.751315	1690.46	12500	9391.44	**10250**
4	0	2250	2250	10	0.683013	1536.78	12500	8537.67	**10250**
5	0	2250	2250	10	0.620921	1397.07	12500	7761.52	**10250**
6	0	2250	2250	10	0.564474	1270.07	12500	7055.92	10250
7	0	2250	2250	10	0.513158	1154.61	12500	6414.48	10250
8	0	2250	2250	10	0.466507	1049.64	12500	5831.34	10250
9	0	2250	2250	10	0.424098	954.22	12500	5301.22	10250
10	0	2250	2250	10	0.385543	867.47	12500	4819.29	10250
11	2500	2250	4750	10	0.350494	1664.85	12500	4381.17	7750
12	0	2250	2250	10	0.318631	716.92	12500	3982.89	10250
13	0	2250	2250	10	0.289664	651.74	12500	3620.80	10250
14	0	2250	2250	10	0.263331	592.50	12500	3291.64	10250
15	0	2250	2250	10	0.239392	538.63	12500	2992.40	10250
16	0	2250	2250	10	0.217629	489.67	12500	2720.36	10250

17	0	2250	2250	10	0.197845	445.15	12500	2473.06	10250
18	0	2250	2250	10	0.179859	404.68	12500	2248.23	10250
19	0	2250	2250	10	0.163508	367.89	12500	2043.85	10250
20	0	2250	2250	10	0.148644	334.45	12500	1858.05	10250
21	2500	2250	4750	10	0.135131	641.87	12500	1689.13	7750
22	0	2250	2250	10	0.122846	276.40	12500	1535.57	10250
23	0	2250	2250	10	0.111678	251.28	12500	1395.98	10250
24	0	2250	2250	10	0.101526	228.43	12500	1269.07	10250
25	0	2250	2250	10	0.092296	207.67	12500	1153.70	10250
26	0	2250	2250	10	0.083905	188.79	12500	1048.82	10250
27	0	2250	2250	10	0.076278	171.62	12500	953.47	10250
28	0	2250	2250	10	0.069343	156.02	12500	866.79	10250
29	0	2250	2250	10	0.063039	141.84	12500	787.99	10250
30	0	2250	2250	10	0.057309	128.94	12500	716.36	10250
31	0	2250	2250	10	0.052099	117.22	12500	651.23	10250
Total			122500			65950.93	385500	116710.80	263000

O& M: Operation and maintenance; DR, DC and DB are the discount rate, discounted costs and discounted benefits.
Benefit –cost ratio = discounted benefits/discounted costs = 116710/65950.93 = 1.77
NPW=discounted benefits -discounted costs = 50760
Payback period= 5 years (bold figures in column 10)
IRR as per calculation with financial function IRR is 25%

ECONOMIC STUDIES OF LAND RECLAMATION

The cost-benefit ratios of alkali land reclamation programs implemented by Central Soil salinity Research Institute, Karnal and the land reclaimed by the farmers have been worked out. The analysis suggests that the benefit-cost ratio even at the farmer's level of adoption is 3 or more than 3 depending upon the component of the subsidy. With 70% subsidy, the B-C ratio is as high as 4.6 (Table 7.9). The B-C ratio could still be higher if the complete rather than a partial package is implemented. The cost of reclamation is generally recovered within 3 years. Recent analysis carried out by Tripathi and Sharma (2011) confirms such results and even at the current prices shows a high IRR of 21.4%. Economics studies carried out by Haryana Land Reclamation and Development Corporation showed that a fair return on investments including repayment of ESI is possible even in the very first year.

Table 7.8 Initial establishment and operational costs to calculate payback period for alkali land reclamation

Particulars	Cost (Rs./ha)	Percentage of total	Returns (Rs./ha)
Ist year			31,600
Bunding and leveling	2,050	1.85	
Tube well installation	8,925	8.07	
Gypsum application	18,650	16.85	
Rice cultivation	14,000	12.66	
Wheat cultivation	13,000	11.75	
2nd year			39,500
Rice cultivation	14,000	-	
Wheat cultivation	13,000	-	
3rd year			42,850
Rice cultivation	14,00υ	-	
Wheat cultivation	13,000	-	
Total (3 years)	**110,626**	**100.00**	**113,750**

Table 7.9 Benefit-cost (B:C) ratio, net present worth (NPW), net benefit investment ratio and payback period of alkali land reclamation technology

Indicators	Subsidy level		
	Without subsidy	**50% subsidy**	**70% subsidy**
As per CSSRI recommendation			
Benefit-cost ratio	4.6	6.0	7.0
Net present worth (10^3 Rs./ha)	66.0	70.4	72.5
Net benefit investment ratio	7.0	11.4	16.7
Payback period (years)	2.0	1.8	1.6
As per farmer's level of adoption			
Benefit-cost ratio	3.0	3.9	4.6
Net present worth (10^3 Rs/ha)	36.7	41.0	43.0
Net benefit investment ratio	3.0	4.6	5.6
Payback period (years)	3.4	2.0	2.6

Datta and Joshi (1991) assessed the financial feasibility of investments on subsurface drainage (Table 7.10). The B-C ratio and the NPW were estimated at 1.26 and Rs. 4,659 per ha. respectively. The IRR of 13.3% was sufficiently higher than the market rate of interest. It may be noted that values for each paramcter are higher under scientific experiments than under farmer's fields mainly because the farmers are unable to implement the full package of land reclamation. The sensitivity analysis by changing the net farm income showed that the drainage for salinity control is a financially feasible proposition.

Table 7.10 Economic parameters of a subsurface drainage project

Criteria	Data sources for analysis	
	Scientific experiments	Farmers' fields
B-C ratio	3.99	1.29
NPW (Rs./ha)	7500	4659
IRR (%)	52.6	13.3

Note: Achievement is based on the yields obtained in the first 3 years

Similar results have been reported from a number of sites with payback period ranging from 3-4 years (Table 7.11, IDNP, 2002). Higher values of NPW and IRR have been obtained in Kerala where reclamation of saline lands was initiated on barren waterlogged saline soils (Fig. 7.1, Mathew *et al.*, 2003). The sensitivity analysis carried out by taking different yields revealed that B:C ratio is 1.0 if rice production increases by 0.41 t/ha (Fig. 7.2).

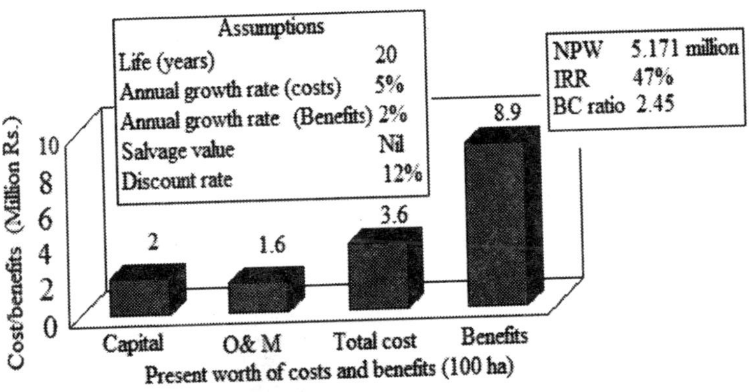

Fig. 7.1 Financial parameters of waterlogged saline land reclamation in Kerala

Table 7.11 Economic parameters of four drainage projects under varying agro-climatic conditions

Particulars	Konaki	Uppugunduru	Segwa	Lakhuwali
B:C ratio	2.8	2.5	1.7	3.2
NPW (Rs./ha)	36, 655	50,872	74.957	51,500
IRR (%)	32	36	58	39
Payback period (Years)	3	3	3	4

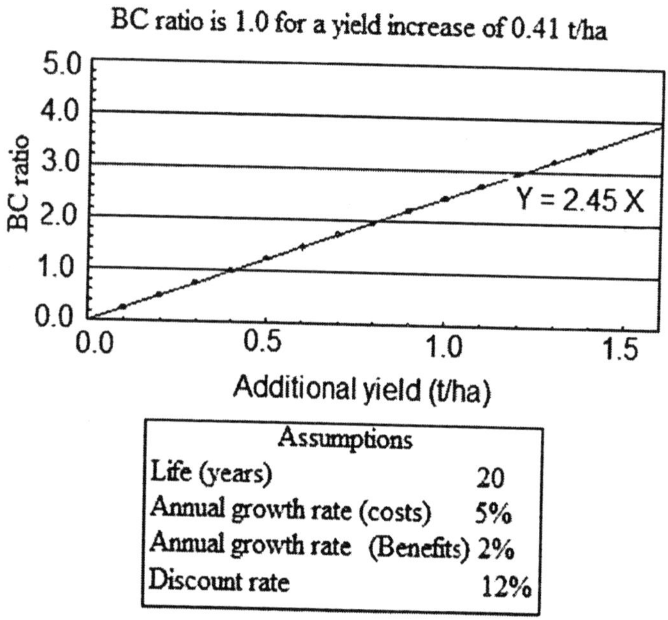

Fig. 7.2 Sensitivity analysis of B:C ratio with respect to yield increase due to drainage

SOCIAL CRITERIA

In many projects, social benefits overweigh the financial benefits. Therefore, such projects are sometimes approved even if these do not stringently satisfy the economic and financial criteria. With emphasis on employment generation, a project even with low BC ratio could be approved provided it generates substantial employment. National and international funding organizations finances many schemes with low BC ratios provided such schemes generate employment, rehabilitate the poor and bridge gender disparity. Some projects are approved if these improve the hygienic or the environmental conditions of an area.

Human Health

Canal irrigation has often been related with increased human health problems due to increased incidences of water borne diseases. While running water could be responsible for this situation in some cases, in others stagnating water could be responsible for such a situation. Increased incidences of malaria have been reported from waterlogged areas as Anolpheles mosquitoes breed in the standing water. Documented evidence of such a situation is a report on migration of people living in and around Amritsar in the first decade of the last century. The vertical drainage scheme designed to halt the migration of people from Amritsar (India) in 1917 could be quoted on this issue. The main aim of the scheme was to improve the general hygienic conditions of the area by lowering the water table. As per IWMI reports, government expenditure on health services could be curtailed in the same order of magnitude as the investments made in the land reclamation programs.

Employment

A social benefit that must be included in any evaluation of a project is its capacity to generate employment opportunities. It has been documented that both alkali land and waterlogged saline land reclamation programs generate huge employment. This alone could be a good ground for the approval of land reclamation projects. While trained engineers will be employed in drainage construction, skilled and unskilled labour would be required to undertake various activities in these projects. At the field level, it will generate work for the farm family/landless people. Capacity to generate employment opportunity at the field level due to reclamation of alkali lands is given in Table 7.12. In the first year, the employment generation is due to land reclamation as well as due to agricultural operations on the reclaimed lands. In subsequent years, it is mainly due to agricultural operations alone. The employment generation potential is still higher in reclamation of waterlogged saline soils with drainage intervention and is related to spacing of laterals (Table 7.13). The employment generation in agricultural operations would be of the same magnitude as in alkali land reclamation.

Table 7.12 Employment generation at field level due to reclamation of 1 ha of barren alkali land

Year	Particulars	Number of		
		Person days	Bullock days	Tractor hours
I	Reclamation	31	10.0	13.3
	Rice	70	1.0	-
	Wheat	47	7.5	3.4
II	Rice	87	6.9	3.0
	Wheat	35	5.6	3.8

Source: Singh (1980)

Table 7.13 Employment generation in saline land reclamation projects involving pipe drainage

Spacing (m)	Installation of laterals and collectors	Construction of		
		Junction boxes	Sump	Total
25	290	5	14	309
50	170	3	14	187
75	112	2	14	128
100	103	2	14	119

Source: Joshi *et al.* (1987)

Inter-sectoral Terms of Trade

Resource mobilization from one sector to another is an indication of strengthening inter-sectoral terms of trade and economic development. Amendment is a major input in any alkali land reclamation program. While land reclamation programs provide opportunities to use low quality gypsum, mining of additional gypsum provides employment opportunities. As the current thinking is to use phospho-gypsum, distillery spent wash and press mud, all byproducts of the industrial sector, it should help the industries to dispose of their byproducts in a fruitful manner. Besides, it would minimize the pollution of soil and water resources.

Total cost involved in constructing a subsurface drainage system can be associated with three different sectors. The share of the industrial sector dominates the total cost of drainage being as high as 59% (Table

7.14). The share of the rural based non-farm resources occupies the second place with 20.8%. The data suggest that land drainage projects should strengthen both the industrial and the agriculture sector.

Table 7.14 Share of different sectors in total cost of drainage

Sector	Share in total cost (%)
Industrial	59.1
Rural based non-farm resources	20.8
Rural labour	20.1
Total	100.0

Source: Joshi *et al.* (1987)

Increase in the Land Value

One of the major secondary benefits of land reclamation is that there is instant enhancement in the land value. The data in Table 7.15 show that all kinds of alkali lands were quoted at higher rates after reclamation although the maximum increase in the value was for the C grade soils, which were lying barren before reclamation. Similar increase in land value was observed for waterlogged saline lands (Fig. 7.3). If this increase in land value is accounted, it alone takes care of the investments made in land reclamation.

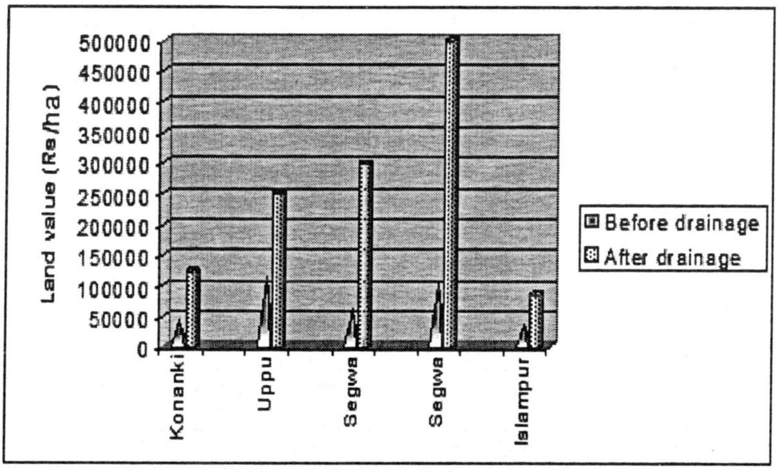

Fig. 7.3 Increase in land value after reclamation at various sites

Table 7.15 Land value before and after reclamation in Uttar Pradesh

Category of land	Land value (Rs./ha)	
	Before reclamation	After reclamation
B+	115,000	170,000
B	55,000	155,000
C	30,000	125,000

Source of Irrigation Water

Subsurface drainage studies in India have shown that drainage besides reclaiming saline soils provides a source of water, which helps to replenish the meager water supplies through the canal system. It has been observed that water quality at some sites is better than the tube well water being used by the farmers. At other sites, the drainage water quality improves with time. Within 2-3 years, drainage effluent can be reused within the reclaimed land to provide supplemental irrigations (Fig. 7.4).

Fig. 7.4 Reuse of drainage water at Kalayat (Haryana)

Ground Water Recharge

As the land reclamation process is initiated, there is perceptible improvement in the infiltration rate and the hydraulic conductivity of alkali lands. As such, it not only reduces the total runoff but also the peak rate of runoff. The water retention capacity of the watershed is also higher because of *bunding* and other on-farm development works. All these together reduce flooding and increase ground water recharge. Thus, a part of the ground water exploited for rice-wheat cultivation in reclaimed alkali lands is compensated by this process. Current technology of storing rainfall in rice fields could further enhance the ground water recharge. It may not be out of place to mention that more than 90% of the rainfall occurring in reclaimed catchments could be stored within the area including the rice fields. Similarly, provision of drainage vastly improves the surface and subsurface components in waterlogged saline lands. Thus, the conservation of rainfall in saline lands vastly alters the salinization-desalinization cycle resulting in reduced salinization and increased desalinization. As such, the reuse of drainage effluent is made possible during winter cropping. Besides solving the disposal related problems, the reuse helps to increase productivity as a result of additional numbers of irrigations.

Enhanced Status of the Poor in Rural Setting

In a rural setting, farmers having barren salt affected lands are ill-treated mainly because of their poor economic status. They are forced to work on someone else's fields where they are forced to do all kinds of works. With reclamation, these farmers become the proud owners of lush green fields. They are able to engage their family members at their own farm and in some cases even provide employment to few landless laborers. With this transformation, they are now able to play an independent role in socio-politico-economic functions of the village. Farmers, who used to migrate from the village, could stay back to take care of their families. It should halt ill treatment of women folk, which might happen in the absence of male members upon migration.

Cost Recovery

Land reclamation is a costly proposition beyond the reach of the farmers. The cost at current prices could be around Rs. 45,000 - 60,000 per ha. Therefore, it is imperative that the government provides initial investments, appropriate subsidy and one window support for other inputs. Although quite appropriate, still questions arise as to 'who will ultimately pay the costs' and 'what portion of the costs should be recovered from the farming community'? In an era when subsidies, are being phased out, one might argue that the whole cost must be recovered from the beneficiary farmers. At the same time, it would be prudent that the recoveries are commensurate with the net benefits. Although in a alkali land reclamation program, it is easier to assess the benefits, the assessment of net benefits in a land drainage project could be a complex process as the total benefits are governed by:

- Elevation of the land in relation to drainage
- Proximity of main drain to the land
- Distance from the natural outlet
- Thoroughness of drainage
- Soil fertility
- Variations in prevailing grades.

Clearly the net benefits in a project would be highly variable. It is therefore, necessary to assess this variability before the costs are apportioned amongst the beneficiaries. Such an exercise might be beyond the reach of human resource available in a project. Therefore, whole area could be divided into blocks of 15-20 ha to assess the benefits by a group of farmers. The other methods of distributing costs amongst the beneficiaries could be:

- Arbitrary assessment
- Uniform distribution on per ha basis
- Assessment on the basis of enhancement in land value
- Assessment by classification, and
- Assessment on the basis of degree based on 100 points scale.

In the last case, a field deriving maximum benefits is assigned an index of 100 points. Other areas with relatively less benefits are awarded points on this basis. These points are then used to distribute the whole cost of the project amongst the beneficiaries.

It is reasonable that general benefits to the public property should be clearly worked out and should not be charged to the farmers. In other words, 100% of the cost should not be recovered from the farmers. Moreover, inhabitants other than the farming community are also benefitted from land drainage. It also strengthens the argument that it would be inappropriate to recover the whole cost from the farming community alone. Therefore, the recovery issue needs to be examined thoroughly to frame some reasonable mechanism and guidelines.

Subsidy

Another way to compensate the farmers is to provide subsidy on land reclamation. There is ample justification for subsidy. It should either be adjusted in the recovery rates or should be granted during the project implementation. It may be noted that currently subsidy is available for the alkali land reclamation as well as for the saline land reclamation programs. The following factors justify the subsidy.

- Most lands have been damaged due to canal irrigation for which farmers are paying the revenue to the state exchequer. They should not be penalized for technical faults in the planning of a project.
- Reclamation projects have many indirect social benefits which will be shared by the whole population of the village. Since the government would also be sharing the benefits (reduced damage to public property and reduced expenditure on health), appropriate subsidy to the farming community is justified.

PROGRESS OF RECLAMATION

Reconciliation of data on area reclaimed by different organizations, gypsum distributed for land reclamation and subsurface drainage installed in different states, it has emerged that 1.85 M ha of barren salt affected soils have so far been reclaimed in India, major reclamation activity being in the states of Punjab, Haryana and Uttar Pradesh. Rajasthan leads in reclamation of saline lands through subsurface drainage followed by Haryana and Maharashtra. Currently, on an average 50,000 ha of salt affected soils are being reclaimed per annum. The massive efforts that have gone in land reclamation has ensured household food security and changed the socio-economic profile of more than 9 million people directly. This has also benefitted a large number of landless labourers indirectly. In spite of the achievements made so far, there is a need to enhance the pace of land reclamation as total salt affected soils currently are at about the same level as reported by Abrol and Bhumbla in 1971. Indirectly, it means that the pace of reclamation and degradation is at about the same level during the last 40 years. The land reclaimed so far and direct and indirect benefits in terms of production, income and employment generation are given in Fig. 7.5. This shows that it has contributed around 15 million tonnes of food grains valued at 13,500 million Rs. It has generated 8.33 million person months of employment (Tripathi and Sharma, 2011).

CONSTRAINTS IN LAND RECLAMATION

Technical and economic feasibility of reclamation of salt affected soils has been proved beyond doubt. In spite of this, the pace of reclamation is not as fast as one would expect it to be. The constraints that hinder the pace of reclamation can be grouped into the following three categories:

- Financial
- Social
- Administrative and political

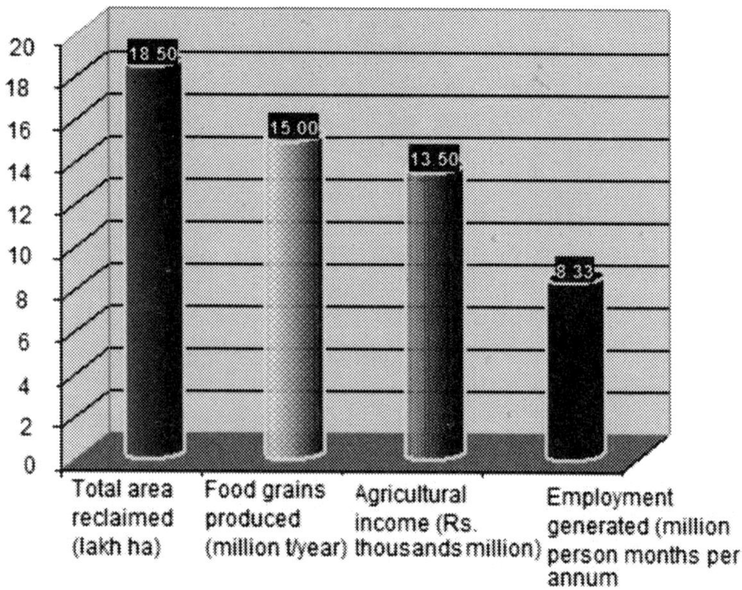

Fig. 7.5 Area reclaimed and some direct and indirect benefits at the national level

Financial Constraints

Any reclamation program can be initiated only if the farmer has substantial funds for initial investment. The farmers with small holdings that too affected by water logging and/or salts do not have sufficient funds to invest on reclamation work. In spite of the loan facilities being extended in some states, farmers are unwilling to take loans because of the tedious and time-consuming nature of the processing of loan applications. To overcome this and other hurdles, it might be useful to associate agriculture development officers of the lead banks in sensitizing the farmers through appropriate training programs. Since, the technologies are more effective and viable if it is implemented as a full package, the issue must be discussed in training programs and appropriate funds for each component should be included in the loan amount.

In most developing countries several projects compete with each other for investment within the limited funds available. It is, therefore, necessary to strike a balance between these conflicting demands, goals to be achieved and the compelling need to increase food production. Since land reclamation is a priority to raise food production and for the general uplift of the rural masses, investments on this activity cannot be neglected, although huge investments might be needed to cover large areas that have gone out of cultivation. It may be noted that besides the primary investments at the project level, secondary investments would be needed in supporting activities. For example, investments are needed in industrial sector to produce drain tiles and envelope material if reclamation of waterlogged saline lands is being contemplated on a large scale. Initial investments would be needed to procure drainage machines and raw material (Table 7.16). There should be no hesitation to release the funds as the gestation period between the investments and benefits realization is small.

Table 7.16 Machinery required for a fleet of 2-trenching machines

Machine/equipment	Numbers	Machine/equipment	Numbers
Trencher	2	Two way radios	9
Tractor and trolleys/wagons	6	Survey equipment sets	2
Backhoes, 4 wheel drive	2	Construction pumps	2
Bulldozer, D-6 equivalent	2	Portable generator	1
Backfill machine	1	Standby generator	1
4-wheel loaders	2	Water truck	1
Field utility equipments	1	Maintenance truck	1
Pick-up trucks	5	Maintenance trailers	2
Mini bus	2	Spare parts	25 % of cost
Laser grade control* units	1	Wrapping Machine	@

*Additional back-up laser units, @ A wrapping machine for wrapping filter around 80 mm lateral pipes. One/two wrapping machine may be provided for several sites rather than to allocate the machine to each site.

Social Constraints

The reclamation technology has been found to have wide acceptance amongst the farmers. No social constraints are visible in any region where these technologies have been adopted. Traditional technologies to scrap the top layers, mixing of sand and FYM application have not paid due dividends. As such, there are no competitive technologies. It seems that lack of extension is a major hindrance. Marginal and small farmers have low level of knowledge regarding salt affected soils and their reclamation. Probably an extensive mobile training network with an agricultural institute in the region as the nodal point might help in spreading the message quickly and effectively.

Administrative and Political Constraints

Realizing the hardships of small and marginal farmers, Governments of Punjab, Haryana and Uttar Pradesh encouraged land reclamation by setting up Land Reclamation Corporations. Initially, they had their own pilot projects to demonstrate the feasibility of the technologies and to identify technological gaps. Subsidy on gypsum was provided and electrical connections to energies tube wells were given on priority. This model was not adopted by other states (Gupta *et al.*, 2011).

A major constraint, with respect to alkali land reclamation is with regard to *Panchayat*/organizational lands. These lands are given to an individual farmer on short-term lease basis. As such, these farmers do not take interest in the reclamation of leased lands. Therefore, a policy to lease out these lands at least on medium term (say 6 years) would be desirable. If it is done then the farmers would be able to get back their investments in about 3 years time and earn additional income for another 3 years. Another limitation is due to land holding acts. Many farmers owning large chunks of alkali lands take no interest in reclamation lest a part of their holding becomes surplus upon reclamation. A suitable strategy to deal with such difficulties needs to be evolved to encourage land reclamation.

In spite of these constraints, the active support provided by many state governments to the alkali land reclamation program has gone well with the farmers. Positive features of this technology that helped to gain farmers support are listed in Table 7.17. On the contrary, waterlogged saline land reclamation program is spreading at a slow pace. In many states, where gravity outlets are available farmers are coming forward to undertake land drainage with their own investments. It is believed that if the constraints as listed in this section and Table 7.17 for reclamation of saline lands are addressed, the technologies would gain further support. Even the *Panchayats* in that event would earn more while turning their white deserts in a lush green landscape.

Table 7.17 Adoption potential of land reclamation technologies

Features of Technologies likely to be adopted easily	Alkali land reclamation technology	Saline land reclamation technology
Direct effects visible	√	√
Technically flawless	√	√
Individual farmer oriented	√	Only if gravity outlet is available
Economically viable under farmers practices and inputs	√	√
Low initial investments	√	√
Simple technology	√	Technical support required
Policy incentives are available	√	√
Institutional support	√	Limited in few states
In line with the existing social norms	√	√

Environmental Impact Assessment

The environment Impact Assessment (EIA) of a large reclamation project comprises of the following main steps:

- Detailed analysis of the useable land losses due to permanent land acquisition by the project components and its economic and social impacts, resettlement and compensation
- Terrestrial and aquatic ecology (Flora and fauna of the direct impact zone)
- Detailed analysis of epidemiology of the water based and water related diseases such as malaria, schistosomiasis and other contagious diseases
- Assessment of types of biocides (insecticides, herbicides, fungicides and rodenticides etc.) currently used as well as their progressive inputs on aquatic ecologies of the recipient water bodies from the increased rate of application following implementation of the project
- Likely impacts associated with construction and operational phases
- The effect of the project on other water use patterns, its quality, availability particularly surroundings of the project area
- Assessment of the socio-economic activities surrounding the proposed project that will have an adverse impact on the project or *vice-versa*

An environmental baseline assessment (EBA) must precede a large land reclamation project. In this assessment existing and potential environmental issues are identified and assessed. Such an assessment should help to finalize an environmental Management Plan (EMP).

The alternative options should include several alternative technologies, alternative designs and phasing of the project. The alternative options should be evaluated economically to arrive at the best option. The main difficulty could be to arrive at appropriate monetary numbers for social and environmental plus and minuses. An essential comparison is between the project and no project option.

The main focus of the environmental impact assessment of a drainage project should revolve around options for the environmentally sound disposal/reuse of drainage effluent since the major impact of a drainage project could be on the stream flow and its quality. Thus, pre and post project discharge of the streams and water quality must be monitored. Post drainage discharge in the surface water bodies and specifically in the main drain is likely to increase. Thus, the carrying capacities of all the drains leading to main drain should be assessed for their sufficiency. Moreover, the drainage effluent in arid and semiarid regions being salty, might affect the quality of water bodies depending upon the quantum of discharge. The water quality both at the upstream and downstream of the project should be measured and assessed. The parameters of interest are: EC, pH, NO_3, PO_4 and trace elements. Both nitrate and phosphorous contribute to eutrophication of waterways. Nitrate being highly soluble and mobile, it would find its way through the soil into the subsurface drainage effluent. Mobility of phosphate being limited it might find its way through the surface drainage system. Since the phosphate promotes the growth of aquatic plants and organisms such as algae, it would have a major impact on flow efficiency and capacity of drains.

The potential harmful levels of trace elements in effluents are widely cited as an environmental issue for drainage projects. Various trace elements such as arsenic, aluminium, lead, manganese, boron, cadmium, selenium may negatively impact the crops/plants, aquatic and terrestrial life. Such contaminants may exist naturally in soils or get deposited by contaminated irrigation water, air pollution or breakdown products from pesticides. While many heavy metals are not readily mobile, irrigation and drainage may provide a mechanism and pathways for contaminants to get transported to receiving waters.

REFERENCES

Abrol, I.P. and Bhumbla, D.R. 1971. Saline and alkali soils in India- their occurrence and management. In: World Soil Series, FAO Report No. 41. 42-51.

Abrol, I.P. and Bhumbla, D.R. 1973. Field studies on salt leaching in a highly saline-sodic soil. Soil Sci. 15:429- 433.

Abrol, I.P. and Bhumbla, D.R. 1979. Crop response to differential gypsum application in a highly sodic soil and tolerance of several crops to exchangeable sodium under field conditions. Soil Sci. 127:79-85.

Abrol, I.P., Dahiya, I.S. and Bhumbla, D.R. 1975. On the method of determining gypsum requirement of soils. Soil Sci. 120: 30-36.

Abrol, I.P., Dargan, K.S. and Bhumbla, D.R. 1973. Reclaiming Alkali Soils. Bull. 2, CSSRI, Karnal.

Abrol, I.P. and Verma, K.S. 1978. A comparative study of the effect of gypsum and pyrites on yield of rice and wheat grown in a highly sodic soil. Proc. FAI-PPCL Sem. Use of Sedimentary Pyrites in Reclamation of Alkali Soils, Lucknow, India. 136a-136c.

Abrol, I.P., Yadav, J.S.P and Massoud, F.I. 1988. Salt Affected Soils and Their Management. FAO Soils Bulletin 39. Food and Agriculture Organization of the United Nations, Rome.

Acharya, C.L. and Abrol, I.P. 1991. Soil Water Relation and Root Water Extraction in Alkali Soils. Tech Bull. 16. CSSRI, Karnal. 96 p.

Acharya, C.L., Sandhu, S.S. and Abrol, I.P. 1979. Effect of exchangeable sodium on the rate and pattern of water uptake by raya (*Brassica juncia*) in the field. Agronomy J. 71: 936-941.

Agarwal, R.R., Yadav, J.S.P. and Gupta, R.N. 1982. Saline and Alkali Soils of India. Indian Council of Agricultural Research, New Delhi. 286 p.

Ahmed, E. 1972. Coastal Geomorphology of India. Orient Longman, New Delhi. 222 p.

AICRP, 2007. Biennial Report, 2005-06. AICRP Management of Salt Affected Soils and Use of Saline Water in Agriculture. CSSRI, Karnal. 171 p.

AICRP, 2009. Biennial Report, 2006-08. AICRP Management of Salt Affected Soils and Use of Saline Water in Agriculture. CSSRI, Karnal. 212 p

Ambast, S.K. and Gupta, S.K. 2007. Design of surface drainage system. In: Agricultural Land Drainage: Reclamation of Waterlogged Saline Lands (Ambast, S.K., Gupta, S.K. and Singh, Gurbachan Eds.) Central Soil Salinity Research Institute, Karnal. 82-91.

Ambast, S.K., Raja, R. Chaudhuri, S.G., Gupta, S.K., Singh, Gurbachan and Yaduvanshi, N.P.S. 2007. Restoring Crop Production in Tsunami Affected Agricultural Lands of A & N Islands: A Success Story. CARI, CSSRI Action Aid Publication. Port Blair. 6 p.

Ambast, S.K., Sen, H.S. and Tyagi, N.K. 1998. Rainwater Management for Multiple Cropping in Sundarbans Delta (WB). Bulletin No 2/98, Regional Research Station, Central Soil Salinity Research Institute, Canning Town, India. 69 p.

Anonymous. 2007. Biotechnology for abiotic stress tolerance. Biotech News. Department of Biotechnology, New Delhi. 11: 9.

Arar, A. 1971. Irrigation and drainage in relation to salinity and waterlogging. Irrigation and Drainage Paper 7. Salinity Seminar, Baghdad, FAO Rome: 86-111.

Ayers, R.S. and Westcot, D.W. 1976. Water Quality for Agriculture. FAO Irrigation and Drainage Paper 29. FAO, Rome.

Ayers, R. S. and Westcot, D.W. 1985. Water Quality for Agriculture. Irrigation and Drainage paper 29, Rev. 1. FAO, Rome. 174 p.

Bains, S.S. and Singh. K.N. 1966. Utilization of solar radiation in desalinization of ridged plant beds on saline soil. Nature. 212: 1391-1392.

Babcock, K.L., Egorov, V.V. and Kovda, V.A. 1973. Chemistry of saline and alkali soils of arid zones. In: Irrigation Drainage and Salinity. Kovda V.A., Berg C. and Hagan R. N. (Eds). Hutchinson-FAO-UNESCO Publication.

Bajwa, M.S., Choudhary, O.P. and Josan, A.S. 1992. Effect of continuous irrigation with sodic and saline sodic water on soil properties and crop yield under cotton-wheat rotation in Northern India. Agricultural Water Management. 22: 345-350.

Bajwa, M.S., Choudhary, O.P. and Josan, A.S. 1998. Sodic waters and their management. In: Agricultural Salinity Management in India (Tyagi and Minhas, Eds.). Central Soil Salinity Research Institute, Karnal. 431-450.

Bajwa, M.S., Hira, G.S. and Singh, N. T. 1983. Effect of sodium and bicarbonate irrigation waters on sodium accumulation and maize and wheat yields in northern India. Irrigation Science. 4: 91-99.

Bandyopadhyay, A.K. and Sen, H.S. 1977. Introduction of a second rice crop and mulching as reclamatory management practices in saline soils of mono-cropped areas of south 24-Parganas, West Bengal. J. Indian Soc. Soil Sci. 25: 326-330.

Batta, R.K. 2011. Micro-irrigation for water conservation and high productivity. In: Salinity Management for Sustainable Agriculture in Canal Commands (Dey & Gupta Eds.). Central Soil Salinity Research Institute, Karnal. 34-42.

Batta, R.K.; Pandey, R.N. and Gupta, S.K. 1980. Studies to improve uniformity of leaching under tile drainage system. Annual Report, CSSRI, Karnal. 83 p.

Bernstein, L. 1961. Tolerance of plants to salinity. Am. Soc. Civil Engg. Proc. 87: 1-12.

Bernstein, L. 1964. Salt Tolerance of Plants. USDA Agri. Inf. Bull. No. 283. USDA, Riverside, California, USA.

Bernstein, L. 1974. Crop growth and salinity. In Drainage for Agriculture. (Ed. J. van Schilfgaarde). Agronomy Monograph, 17. Am. Soc. Agronomy, Washington 39-54.

Bernstein, L. and Francois, L.E. 1973. Leaching requirement studies: sensitivity of alfalfa to salinity of irrigation and drainage waters. Soil Sci. Soc. Am. Proc. 37:931-43.

Beyce, O. 1972. Drainage of salty soils. Irrigation and Drainage Paper 16: FAO Rome: 63-67.

Bhatt, M.N., Goel, K.N. and Uprety, M.C. 1978. Use of pyrites in reclamation of alkali soils in Uttar Pradesh. Proc. FAI-PPCL Sem. Use of Sedimentary pyrites in Reclamation of alkali soils, Lucknow, India. 107-114.

Bhumbla, D.R., Kanwar, J.S., Mahajan, K.K. and Singh, B. 1964. Effect of irrigation with different sodium and salinity hazards on the growth of crops and the properties of soils. Proc. Symp. Problems of Indian Arid Zone, Jodhpur.

Bos, M.G. 1977. Arid land irrigation in developing countries- environmental problems and effects. (Ed. Worthington E.B.). Pergamon Press, Oxford: 193-204.

Boumans, J.H. and van Der Molen, W.H. 1964. Out watering- behoefts can bevolede gronden in verband met hun zouthuishouding. Landis Tydrschn ft. 76:880-887.

Bouwer, H. 1969. salt balance, irrigation efficiency and drainage design. ASCE Proc. 95 (IR1): 121-144.

Bower, C.A. 1961. Prediction of the effect of irrigation water on soil. Proc. Tehran Symp. on Salinity Problems in Arid Zones. 1-8.

Bower C.A. and Massland, M. 1963. Sodium hazard of Punjab groundwaters. West Pakistan Engg. Congress Proc. 49-61.

Burman, D., Bandyopadhyay, B.K. and Bal, A.R. 2010. Annual Report 2009-10. Central Soil Salinity Research Institute, Karnal. 99-100.

Burman, D., Bandyopadhyay, B.K. and Bal, A.R. 2011. Annual Report 2010-11. Central Soil Salinity Research Institute, Karnal. 95-96.

Burns, I.G. 1975. An equation to predict the leaching of surface -applied nitrate. J. Agric. Sci. Camb. 85:443-454.

Burns, I.G. 1976. Equations to predict the leaching of nitrate incorporated to a known depth or uniformally distributed throughout soil profile. J. Agric. Sci. Camb. 86:305-311.

Burley, J. 1980. Selection of species of fuel wood plantation. Comm. for Rev. 59:133-147.

Canada, Department of Agriculture 1977. Management of Saline Soils. Pub. 1624: 31 p.

Carter, D.L. and Fanning, C.D. 1964. Combining surface mulches and periodic water applications for reclaiming saline soils. Soil Sci. Soc. Amer. Proc. 28: 564-567.

Carter, D.L. and Robbins, C.W. 1978. Salt outflows from new and old irrigated lands. Soil Sci. Soc. Am. J. 42:627- 632.

Chauhan, R.K. S. 1978. Note on comparative study of pyrites and gypsum as soil amendments in alkali soils. Proc. FAI-PPCL Sem. Use of Sedimentary pyrites in Reclamation of alkali soils, Lucknow, India. 127-129.

Chavan, K.N., Chavan, A.S. and Kadrekar, S.B. 1981. Factors affecting salinity status of coastal saline soils of Maharashtra. Proc. Sec. Conf. on Problems and Management of Coastal Saline Soils. CSSRI Regional Res. Stat. West Bengal. 35-42 (Unpublished).

Chawla, K.L. 1985. Salt and Water Transport under High Water Table Conditions. Unpublished Ph.D. thesis. Kurukshetra University, Kurukshetra.

Chhabra, R. 1996. Soil Salinity and Water Quality. Oxford & IBH Publication Pvt. Ltd. New Delhi.

Chhabra, R and Abrol, I.P. 1977. Reclaiming effect of rice grown in sodic soil. *Soil Sci.* 124: 49-55.

Chhabra, R. Singh, N.T. and Hansen, A. 1989. Interaction between type of amendment and P availability. Ann. Rep. Central Soil Salinity Research Institute, Karnal. 38.

Chhabra, R. and Thakur N.P. 1998a. Bio-drainage using trees to control water logging and secondary salinization in canal irrigated areas. Proceedings National Conference on Salinity Management in Agriculture. Central Soil Salinity Research Institute, Karnal, India.

Chhabra, R. and Thakur, N.P. 1998b. Lysimeter study on the use of bio-drainage to control water logging and secondary salinization in (Canal) irrigated arid/semi-arid environment. Irrigation and Drainage Systems. 12: 265-288.

CSSRI. 2004. Reclamation and Management of Salt Affected Soils. Central Soil Salinity Research Institute (CSSRI), Karnal. 160 p.

CSSRI. 1997. Vision-2020. CSSRI Perspective Plan. Indian Council of Agricultural Research. Central Soil Salinity Research Institute. Karnal, India. 95 p.

CSSRI, 2004. Pre and post Narmada canal irrigation effect on soil health and ground water quality under different cropping systems in bara tract of Gujarat. Annual Report -2003-04. CSSRI, Karnal. 81.

CSSRI. 2005. Evaluation of some halophytes for saline agriculture on salt affected black soils of Gujarat. Annual Report -2004-05, CSSRI, Karnal. 82.

CWC. 2010. Water Statistics. Central Water Commission (CWC). Ministry of Water Resources, GoI, New Delhi.

Dagar, J.C. 2005. Ecology, management and utilization of halophytes. Bulletin of the National Institute of Ecology. 15: 81-97.

Dagar, J.C. 2011. Bio-drainage for amelioration of water logging. In: Salinity Management for Sustainable Agriculture in Canal command (Dey and Gupta Eds.). CADA, Haryana and CSSRI, Karnal. 54-58.

Dagar, J.C., Singh, G. and Singh, N.T. 2001 Evaluation of forest and fruit trees used for rehabilitation of semi-arid alkali/sodic soils in India. Arid Land Research and Management, 15: 115-133.

Dagar, J.C. and Singh, N.T. 1994. Agro-forestry options in the reclamation of problem soils. In Trees and Tree Farming (Thampan, P.K. Ed.). Peekay Tree Crops Development Foundation, Cochin (India): 65-103.

Dahiya, I.S., Malik, R.S. and Singh, M. 1981. Field studies on leaching behaviour of highly saline-sodic soil under two modes of water application in presence of crops. J. Agric. Sci. Camb. 97:383-389.

Dahiya, I.S., Singh, M. and Hajrasuliha, S. 1980. Simultaneous transport of surface -applied salts and water through unsaturated soils as affected by infiltration, redistribution and evaporation. Soil Sci. Soc. Am. J. 44:223.

Dandekar, C.B. and Chougule, B.A. 2010. Drainage of irrigated lands. Proc. 9th Int. Drainage Symposium, CIGR and CSBE, 13-16 June, 2010. Canada.

Dargan K.S. and Chillar R.K. 1973. Technique for sugar beet and vegetables in saline alkali soils. Indian Farming. 22: 13-16.

Dargan K.S., Gaul B.L. and Abrol I.P. 1973. Growing sugarcane in saline-alkali soils. Indian Farming. 23:10-19.

Dargan, K.S., Gaul, B.L. and Abrol, I.P. 1976. Effect of gypsum, farmyard manure and method of sowing on the yield of sugarcane in a sodic soil. J. Agronomy. 21:349-353.

Datta, K.K. and de Jong, C. 1991. The effect of subsurface drainage on the farm economy. Report Int. Inst. Land Reclamation and Improvement. Wageningen, The Netherlands.

Datta, K.K. and Joshi, P.K. 1991. Cost and benefits of subsurface drainage for the control of soil salinity and water logging. Better Farming in Salt Affected Soils No. 20. CSSRI, Karnal.

Dhir, R.P. 1977. Saline water, their potentiality as a source of irrigation: In: Desertification and its Control (Anonymous). ICAR, New Delhi. 130-141.

Dhir, R.P. and Bhatia, O.P. 1975. Use of saline waters in agriculture: I description of the system. Ann. Arid Zone. 14: 206-211.

Dhruva Narayana, V.V. 1980. Reclaiming Alkali Soils-Engineering Aspects. Bulletin No. 6. Central Soil Salinity Research Institute, Karnal. 61 p.

Dieleman, P.J. 1963. Reclamation of Salt Affected Soils in Iraq. Voenman, Wageningen. 175-190.

Dumm, L.D., 1954. Drain spacing formula, new formula for determining depth and spacing of subsurface drains in irrigated lands. Agricultural Engineering, American Soc. of Agricultural Engineers. 35: 726-730.

Dumm, L.D., 1964. Transient flow concept in subsurface drainage, its validity and use. Transaction of ASAE, 7: 142-146, 151.

Elgabaly, M.M. 1971. Reclamation and management of salt affected soils. Salinity Seminar, Baghdad. FAO Irrigation and Drainage Paper 7:50-79.

Eaton, F.M. 1950. Significance of carbonate in irrigation water. Soil Sci. 69: 123-133.

Fertilizer Association of India (FAI). 1985. Soils of India and Their Management. Fertilizer Association of India, New Delhi. 445 p.

Finkel, H.J. 1983. Handbook of Irrigation Technology. Vol. II, CRC Press. Florida, USA. 223 p.

Fortier, S. and Scobey, F.C. 1926. Possible canal velocities. Trans. Am. Soc. Civil Engineering. 89: 940-984.

Gardner, W.R. and Fireman, M. 1958. Laboratory studies of evaporation from soil columns in presence of a water table. J. Soil Sci. 85: 244-249.

Garland, K.R. 1966. Reclamation of waterlogged saline lands in Northern Victoria, Australia - A review of recent field experiments. 6th Cong. Irrigation and Drainage. 19: 147- 157.

Gupta, I.C. 1972. Note on the relationship between pH of saturated soil paste and 1:2 soil water suspension. Indian J. agric. Sci. 42: 962-63.

Gupta, I.C. 1984. Reassessment of irrigation water quality criteria and standards. Current Agriculture. 7:150-155.

Gupta, I.C. 1984b. Effect of boron, lithium and fluorine in irrigation water on plant growth. Ann. Rep. CSSRI, Karnal.

Gupta, I.C. and Abichandani, C.T. 1970. Seasonal variations in the salt composition of some saline water irrigated soils of western Rajasthan. J. Indian Soc. Soil Sci. 18: 428-35.

Gupta, I.C. and Bandyopadhya, A.K. 1975. Effect of heating in reclamation of a salt affected soil. Proc. Int. Symp. New Developments in the Field of Salt Affected Soils. Cairo, 1972. 525-528.

Gupta, I.C., Bohra, R.K. and Abichandani, C.T. 1969. Relationships between certain indices of saline water irrigated soils. *Ann. Arid Zone* 8: 27-32.

Gupta, I.C. and Gupta, S.K. 2002. Use of Saline Water in Agriculture. Scientific Publishers, Jodhpur. 302 p

Gupta, I.C., Gupta, S.K. and Sharma, D.P. 2007. Alkali Wastelands, Environment and Reclamation. MD Publications Pvt. Ltd. New Delhi. 411 p.

Gupta, I.C., Yaduvanshi, N.P.S. and Gupta, S.K. 2012. Standard Methods for Analysis of Soil Plant and Water. Scientific Publishers (India), Jodhpur. 224p.

Gupta, R.K. and Abrol, I.P. 1990. Salt affected soils: their reclamation and management for crop production. In: Lal R and Stewart BA (Eds). Advances in Soil Science: III- Soil Degradation, Springer-Verlag, New York. 223-288.

Gupta, S.K. 1981. Evaluation of salt and water balance in saline soils without provision of drains. Annual Rep. CSSRI, Karnal.112-114.

Gupta, S.K. 1982. Evaluating solute movement parameters in one dimensional flow through soils. Unpublished Ph.D. Thesis. J.N.T. University, Hyderabad

Gupta S.K. 1983. Evaluation of leaching and resalinisation behaviour of saline soils in relation to crop growth. Annual Rep. CSSRI: 114-115.

Gupta S.K. 1985. Dynamics of salts in saline water irrigated soils. J. Institution of Engineers (India) 41: 17-22.

Gupta, S.K. 1985. Salt Movement and Dispersion- convection Equation: Material Balance Concept. Tech. Bull. No. 9. Central Soil Salinity Research Institute, Karnal. 48 p.

Gupta S.K. 1985. Subsurface drainage for waterlogged saline soils. Irrigation and Power Journal. 42: 335-344.

Gupta, S.K. 1986. Salinisation-desalinisation in saline water irrigated soils. Proc. Seminar on Problems of Water Logging and Salinity in Alluvial Soils. New Delhi. 12.01-12.16.

Gupta, S.K. 1992. Leaching of Salt Affected Soils. Bulletin No.17, CSSRI, Karnal. 89 p.

Gupta, S.K. 2005. A critical analysis of bio-drainage with special reference to Indian experiences. Souvenir: All India Seminar Reclamation of Waterlogged Saline Soils through Drainage. Institution of Engineers (Kota Chapter), Kota. 81-96.

Gupta, S.K. 2011. Case studies on subsurface drainage systems and guidelines for design and evaluation. In: Salinity Management for Sustainable Agriculture in Canal Commands (Dey, P. and Gupta, S.K. Eds.). Central Soil Salinity Research Institute, Karnal. 11-22.

Gupta, S. K., Ambast, S. K., Singh, Gurbachan, Yaduvanshi, N.P.S., Ghoshal Chaudhuri S. and Raja, R. 2006. Strategies for Improved Agriculture in Tsunami Affected Andaman and Nicobar Islands and Maldives. Central Soil Salinity Research Institute, Karnal, Central Agricultural Research Institute, Port Blair and ActionAid, Port Blair. 88 p.

Gupta, S.K. and Gupta, I.C. 1987. Management of Saline Soils and Water. Oxford and IBH Publishers Co. Pvt. Ltd. New Delhi. 339 p.

Gupta, S.K. and Khosla, B.K. 1996. Salinity control in the root zone of irrigated soils. Proc. Workshop Waterlogging and Soil Salinity in Irrigated Agriculture. CBIP, New Delhi and CSSRI, Karnal. 163-180.

Gupta, S.K. and Pandey, R.N. 1979. A simplified equation for reactive solute transport in soil. J. agric. Sci. Camb. 92: 587-589.

Gupta, S.K. and Pandey, R.S. 1983. A mathematical model for evaluating leaching of saline soils when saline and fresh water are used in conjunction. Indian J. Agric. Sci. 53: 701-706.

Gupta, S.K. and Ramadurgaiah, D. 1984. Mathematical model for leaching and its parameter identification under laboratory and field conditions. J. Institution of Engineers (India). 64:68-71.

Gupta, S.K., Sharma, D.P. and Swarup, A. 2004. Relative tolerance of crops to surface stagnation. J. Agril. Engg. 41: 44-48.

Gupta, S.K., Sharma, D.P., Tyagi, N.K. and Dubey, S.K. 2002. *Lavaniya-Kshariya Mirdaon ka Sudhar aur Parbandh*. CSSRI, Karnal. 153 p.

Gupta, S.K. and Sharma, S.K. 1990. Response of crops to high exchangeable sodium percentage. Irrigation Science 11:173-179.

Gupta, S.K., Tejwani, K.G. and Babu, R. 1971. Drainage coefficients for surface drainage of agricultural lands for different parts of the country. J. Irrig. Power. 28: 53-60.

Gupta, S.K. and Yadav, R.N. 1988. Note on leaching requirement for salt balance in irrigated lands. Current Agriculture. 12:101-104.

Gururaja Rao, G., Nayak, A.K. and Chinchmatapure, A.R. 2003. *Salvadora persica* : A Life Support Species for Salt Affected Black Soils. T. Bull. 1/2003. Central Soil Salinity Research Institute, RRS, Bharuch. 44 p.

Gururaja Rao, G., Nayak, A.K., Chinchmalatpure, A.R., Singh, R. and Tyagi, N.K. 2001. Resource Characterization and Management options for Salt Affected Black Soils of Agro-ecological Region V of Gujarat State. Technical Bulletin 1/2001, Central Soil Salinity Research Institute, Regional Research Station, Anand, Gujarat. 83 p.

Gururaja Rao, G., Singh, R. and Bhargava, G.P. 1993. Vegetation Types in Mahi Right Bank Canal Command, Ukai-Kakrapar Command and Bhal region of Gujarat. Tech. bull. 20. Central Soil Salinity Research Institute, RRS, Anand. 52 p.

Haroon, M.A.R. and Bose, S.C.M. 2004. Use of distillery spent wash for alkali soil reclamation, treated distillery effluent for ferti-irrigation of crops. Indian Farming. 53: 48- 51.

Hoffman, G.J. 1980. Guidelines for reclamation of salt affected soils. Proc. Inter-American Salinity and Water Management. Tech. Conf. Juarez, Mexico. 49-64.

Hoffman, G.J. and van Genuchten M.Th. 1983. Soil properties and efficient water use: water management for salinity control. In: Limitations of Efficient Water Use in Crop Production. Am. Soc. Agronomy, Madison, USA.

Hooghoudt, S.B. 1940. Bijdragen tot dekennis van enige naturrk undige grootheden van de ground. Versi Land bouwk Onderz. 7: 449 p.

Hooja, R., Srinivas, V. and Sharma, G. 1994. Water logging and possible remedial measures in the Indira Gandhi canal command area development project. Nat. Seminar on Reclamation and Management of Waterlogged Saline Soils. CSSRI, Karnal.

Hoon, R.C. 1955. Land Reclamation. Publication of Central Board of Irrigation, India. 2: 160-178.

IDNP. 2002. Recommendations on Water Logging and Salinity Control based on Pilot Area Drainage Research. ICAR/CSSRI/Alterra ILRI Indo-Dutch Network Publication. 100 p.

Inglish, G.C. and Gokhale, V.K. 1927. Report on Land Drainage Experiment at Baramati, Technical Paper No. 24, Bombay Public Works Department, Bombay.

Iyengar, E.R.R. and Kurien, T. 1971. Evaluation of seawater tolerance of crop plants: Response of bajra and wheat to seawater salinity. Indian Journal of Agricultural Research. 5:249-255.

Jain T.C., Saxena G.S., Jain K.C. and Totawat K.L. 1976. Effect of crop rotation, farm yard manure and the quantity of saline irrigation water on the yield of wheat and physico-chemical properties of the soil. Indian J. Agric. Res. 10:1-9.

Jena, K.C. and Mishra, M.R. 2011. Effect of global warming and climate change on coastal zones and sea level. Orissa Review. September – October. 82-94.

Joshi, P.K., Singh, O.P., Rao, K.V.G.K. and Singh, K.N. 1987. Subsurface drainage for salinity control- an economic analysis. Indian J. Agric. Econ. 42: 198-206.

Jury, W.A., Jarrell, W.M. and Devitt, D. 1979. Reclamation of saline-sodic soil by leaching. Soil Sci. Soc. Am. J. 43:1100- 1106.

Kanwar, J.S. 1962. Reclamation of saline-alkali soils. Proc. Symp. Salinity and Alkalinity Soil Problems. I.A.R.I., New Delhi. 54-60.

Kelley, W.P. 1967. Use of saline irrigation water. Soil Sci. 95:385-399.

Keren, R. 2000. Salinity *In* Handbook of Soil Science (Ed. M.L. Sumner). CRC Press, New York, G1-G25.

Khosla, B.K., Dargan, K.S., Abrol, I.P. and Bhumbla, D.R. 1973. Effect of depth of mixing gypsum on soil properties and yield of barley, rice and wheat grown in a saline-sodic soil. Indian J. Agric. Sci. 43:1024-1031.

Khosla, B.K.; Gupta, R.K. and Abrol, I.P. 1979. Salt leaching and the effect of gypsum application in a saline-sodic soil. Agric. Water Mgmt. 2: 193- 202.

Kirkham, Don 1949. Flow of ponded water into drain tubes in soil overlying an impervious layer. Trans. Am. Geophys. Union. 30: 369- 385.

Krepkii, I.S. 1981. Growing trees on saline Chestnut soils of northern Kazakhstan. Lesnoe Khozyaistvo. 1:27-29.

Kumar, A. 1988. Performance of forage grasses in saline soils. Indian J. Agron. 33: 26-30.

Kumar, A. and Abrol, I.P. 1986. Grasses in Alkali Soils. Bulletin No. 11. Central Soil Salinity Research Institute, Karnal. 95 p.

Kumar, R. and Singh, J. 1994. Drainage system for ground water management. Nat. Seminar on Reclamation and Management of Waterlogged Saline Soils. CSSRI, Karnal.

Lal, Cheddi. 2005. Develop design criteria of interceptor drain for preventing/ minimizing water logging in areas along major canals. Progress during 2004-05 (Personal communication)

Lefellar, P.A. and Sharma, R.P. 1977. Leaching of a highly saline-sodic soil. J. of Hydrology. 32: 203-218.

Leather, J.W. 1914. Investigations on *Usar* Land in the United Provinces. Govt. Press, Allahabad.

Lodha, Vandana. 2007. Personal communication. CSSRI, Karnal.

Maas E.V. 1980. Crop tolerance to saline soil and water. In: Prospectus for Bio-saline Research. Proc. US–Pak Biosaline Res. Workshop, Karachi, Pakistan: 205- 211.

Maas, E.V. 1986. Salt Tolerance of plants. App. Agric. Res. 1: 12-26.

Maas E.V. and Hoffman G.J. 1977. Crop salt tolerance: current assessment. J. Irrig. Drainage Div. 103:115- 134.

Manchanda, H.R., Khosla, B.K., Sharma, D.R., Minhas, P.S., Chawla, K.L., Gupta, R.K. and Singh, N.T. 1989. Irrigation and Crop Management with Brackish Waters. Bulletin Better Farming Series No. 16, Central Soil Salinity Research Institute, Karnal. 14 p.

Manjunatha, M.V., Hebbara, M., Patil, S.G., Kuligod, V.B. and Minhas, P.S. 2000. Effect of tree plantation in control of canal seepage and shallow water table in saline Vertisols. National Seminar Environmental Pollution and Management, Dharwad.

Mathew, E.K., Nair, M., Raju, T.D. and Jai Kumaran, U. 2003. Drainage Digest: A Report. AICRP on Agricultural Drainage. Kerala Agricultural University, Thrissur, Kerala. India. 200 p.

Mehta, M.L. 1940. The Formation and Reclamation of *Thur* Land in the Punjab. Res. Publ. Punjab Irrig. Res. Inst. 3:14 p.

Minhas, P. S. 2005. Personal discussions. CSSRI, Karnal.

Minhas, P.S. and Gupta, R.K. 1992. Quality of Irrigation Water: Assessment and Management. Indian Council of Agricultural Research, New Delhi. 123 p.

Minhas, P.S. and Sharma, D.R. 2006. Predictability of existing indices and an alternative coefficient for estimating sodicity build-up using adj R_{Na} and permissible limits for crops grown on soils irrigated with waters having residual alkalinity. J. Ind. Soc. Soil Sci. 54 : 331-338.

Mishra, B. 1996. Highlights of Research on Crops and Varieties for Salt Affected Soils. Research Bulletin. Central Soil Salinity Research Institute, Karnal. 28 p.

Mishra, V.K., Sharma, D.K., Singh, Y.P., Verma, C.L., Damodaran, T., Bhardwaj, A.K. Singh, A.K., Jha, S.K. and Singh, C.S. 2012. CSSRI, Regional Research Station, Lucknow at a Glance. CSSRI, Regional Research Station, Lucknow. 14 p.

Misra, K.L., Porwal, A.K. and Gupta, S.K. 1990. Assessment of crop tolerance to saline irrigation waters. *Bhu Jal* News. 54: 35-40.

Mohan, Lal, Arora, V.K. and Gupta, S.K. 2012. Application and performance of geo-synthetic filter materials as drain envelope in land reclamation in Haryana, Indian Journal of Agricultural Engineering (Accepted).

Mondal, A.K., Obi Reddy, G.P. and Ravisankar, T. 2011. Digital database of salt affected soils in India using Geographic Information System. Journal of Soil Salinity and Water Quality. 3:16-29.

Muhawi, D. 2003. Seawater Farms: Bringing Life from Deserts and Seawater. In http://www.ecoworld.com/Home/articles2.cfm?TID=336

Murthy, R.S. 1982. Benchmark Soils of India. National Bureau of Soil Survey and Land Use Planning, Nagpur. Indian Council of Agricultural Research, New Delhi. 374 p.

NABARD 1998. *Casurina* – A Nitrogen Fixing Tree for Wastelands. National Bank for Agriculture and Rural Development, Mumbai. 18 p.

Nanda, S.S.K. and Rai, M. 1979. A note on the problems of saline lands in Orissa and their reclamation and management. Proc. 1st Conf. on "Problems and Management of Coastal Saline Lands". CSSRI Res. Station, Canning (W.B.)

National Commission on Agriculture (NCA). 1976. Report of the Commission : Climate and Agriculture (Part-IV). Ministry of Agriculture and Irrigation. Govt. of India, New Delhi.

NAS, 1990. Saline Agriculture: Salt-Tolerant Plants for Developing Countries. National Academy of Sciences. National Research Council, Washington, D.C. 143 p.

Nayak, A.K., Sharma, D.K., Mishra, V.K., Jha, S.K., Singh, C.S., Shahabuddin, Md., Srivastava, M., Kale, S.P. and Singh, G.B. 2011. Phosphogypsum : An Alternate Chemical Amendments for Sodic Lands. Tech. Bull. 1/2011. CSSRI, RRS, Lucknow, 42 p.

Nayak, A.K., Sharma, D.K., Mishra, V.K., Minhas, P.S. and Verma, C.L. 2008. Reclamation of saline sodic soil under a rice wheat system by horizontal surface flushing. Soil Use and Management. 24: 337-343.

NEDECO, 1968. Reports on Bhal Reclamation Scheme of State of Gujarat, India. Vol. I to IV. Grontmij n.v. de Bilt, the Hague, Netherlands.

Nielsen, D.R., Biggar, J.W. and Luthin, J.N. 1966. Desalinization of soils under controlled unsaturated flow conditions. 6th Int. Cong. Irrigation and Drainage. New Delhi II: 15-24.

Oster, J.D.; Willardson, L.S. and Hoffman, G.J. 1972. Sprinkling and ponding techniques for reclaiming saline soils. Trans ASAE. 15: 1115-1117.

Paliwal, K.V. and Yadav, B.R. 1976. Irrigation Water Quality and Crop Management in the Union Territory of Delhi. Res. Bull. 9. IARI, New Delhi. 170 p.

Pandey, R.N. and Gupta, S.K. 1978. Equations to predict leaching of soluble salts in saline soils. J. Agric. Sci. Cambridge. 91: 131-134.

Pandey, R.N. and Gupta, S.K. 1981. Salt balance and leaching efficiency in farm irrigation systems. J. Ind. Soc. of Soil Sci. 29: 50-54.

Pandey S.L. and Sinha A.K. 1972. Drainage of agricultural lands. Indian Farming. 22: 56-58.

Parur, H.G. 1971. Review of research work on soil salinity and future lines of work for different zones. In: Soil and Water Research in India: Retrospect's and Prospects. Tech. Bull. No. 22. Ministry of Agriculture, New Delhi.

Patil, B.N., Patil, S.G., Hebbara, M., Manjunatha, M.V., Gupta, R.K. and Minhas, P.S. 2005. Bio-ameliorative role of tree species in salt-affected Vertisols of India. Journal of Tropical Forest Science. 17: 346-354.

Pearson, G.A. 1959. Factors influencing salinity of submerged soil and growth of caloro rice. Soil Sci. 87:198-206.

Poonia, S.R., Jhorar, L.R., Nath J. and Khanna, S.S. 1974. Effect of quality of irrigation water, leaching level and farm yard manure on performance of wheat and pearl millet. Indian J. Agric. Sci. 44: 854-859.

Prichard, T.L., Hoffman, G.J. and Oster, J.D. 1985. Reclamation of saline organic soil. Irrigation Sci. 6:211-220.

Publication Division 2011. India-2011. Ministry of Information and Broadcasting. Government of India, New Delhi. 1233 p.

Puntamkar, S.S., Mehta P.C. and Seth S.P. 1971. Note on the inducement of salt resistance in two wheat varieties by pre-soaking with different salts of varying concentrations. Indian J. Agric. Sci. 41: 717-718.

Raghava, Rao, K.V., Raju, T.S. and Ramesan, V. 1969. An estimate of ground water potential of India- first approximation. Soil Water Management Symposium, Hisar. 6-16.

Raghu Babu, M. Prasad, P.R.K., Ravinder Babu, P., Lakshmi, G.V. and Gupta, S.K. 2009. Technologies for Skimming Fresh water in Coastal Sandy Soils. ANG Ranga Agricultural University, AICRP Centre, Bapatala. 44 p.

Raj, M. and Nath, J. 1984. Estimation of water requirement for reclaiming saline soils. J. Indian Soc. Soil Sci. 32:249-252.

RAJAD, 1995. Analysis of Subsurface Drainage Criteria. Rajasthan Agricultural Drainage Research Project, Kota (Rajasthan).

Ram Sewa, Rao, K.V.G.K. and Visvanatha, N.A. 2000. Impact of subsurface drainage in management of saline soils of the Chambal command. 8th ICID Int. Drainage Workshop, New Delhi.

Ranade, D.H. 1984. Leaching phenomenon in sandy loam soils. Unpublished M. Tech. Thesis, I.I.T. Kharagpur.

Rao, K.V.G.K. and Dhruva Narayana, V.V. 1979. Hydrology and drainage conditions of coastal saline soils. Proc. 1st Conf. on Problems and Management of Coastal Saline Soils. CSSRI, Res. Stn. Canning (Unpublished)

Rao, K.V.G.K. and Leeds-Harrison, P.B. 1990. Water management strategies for improving leaching in subsurface drainage. In Management of Irrigation Systems (Tyagi *et al.*, Eds.). Proc. Nat. Seminar Irrigation Management, Karnal: 221-232.

Rao, K.V.G.K., Singh, O.P., Gupta, R.K., Kamra, S.K., Pandey, R.S., Kumbhare, P.S. and Abrol, I.P. 1986. Drainage Investigations for Salinity Control in Haryana. Central Soil Salinity Research Institute, Karnal, Bull. No. 10: 95 p.

Rao, K.V.R., Ravi, K. and Singh, R. 2009a. Mole drainage to enhance soybean production in waterlogged Vertisols. J. of Agricultural Engineering. 46: 54-58.

Rao, K.V.R., Ravi, K. and Singh, R. 2009b. Mole drainage, a viable alternate to pipe drainage- a case study. 60th Int. Executive Council and 5th Asian Regional Conference, New Delhi (www.tndholkia.com/data/delhi/2.c.d.)

Ravindra Babu, P., Raghu Babu, M. Prasad, P.R.K., Lakshmi, G.V. and Gupta, S.K. 2010. Subsurface Drainage for Reclamation of Waterlogged Saline lands in Krishna Western Delta. ANG Ranga Agricultural University, AICRP Centre, Bapatala. 31p.

Rayachaudhari, S.P. 1962. Land Resources of India. Planning Commission. New Delhi.

Reeve, R.C. 1957. The relation of salinity to irrigation and drainage requirements. 3rd Congr. on Irrig. and Drain. 5: 10.175- 10.187.

Reeve, R.C. and Doering, E.J. 1966. The high-salt-water dilution method for reclaiming sodic soils. Soil Sci. Soc. Amer. Proc. 30:498-504.

Reeve, R.C., Pillsbury, A.F. and Wilcox, L.V. 1955. Reclamation of a saline and high boron soil in the Coachella Valley of California. Hilgardia. 24: 69-91.

Rhoades, J.D. 1968. Mineral weathering correction for estimating sodium hazard of irrigation waters. Proc. Soil Sci. Soc. Am. 32:648-652.

Rhoades, J.D. 1974. Drainage for salinity control. In: Drainage for Agriculture. (Ed. J. van Schilfgaarde). Agronomy Monograph 17. Am. Soc. of Agronomy. Wisconsin, U.S.A.: 433-461.

Ritzema, H.P. 1994. Drainage Principles and Applications. ILRI Pub. 16. International Institute for Land Reclamation and Management (ILRI), Wageningen, The Netherlands. 1125 p.

Sahu, R.K., Chavan, K.N. and Kadrekar, S.B. 1981. Some hydrological observations of coastal saline soils of Konkan. Proc. Sec. Conf. on Problems and Management of Coastal Saline Soils. CSSRI Regional Res. Stat. West Bengal. 199-208 (Unpublished).

Schilfgaarde, van. J., Bernstein, L., Rhoades, J.D. and Rawlins, S.L. 1974. Irrigation management for salt control. J. Irrigation Drainage Div. ASCE. 100: 321- 338.

Schilfgaarde, van. J. 1976. Water management and salinity. Prognosis of Salinity and Alkalinity. Soils. Bull. 31. FAO Rome. 53-68.

Schoonover, W.R. 1952. Examination of Soils for Alkali. University of California, Extension Service, Berkley, California, USA.

Sehgal, J.C. and Abrol, I.P. 1994. Soil Degradation in India: Status and the Impact. Oxford and IBH Publ. Co. Pvt. Ltd., New Delhi

Sen, H.S. 2011. Management of coastal salt affected soils. A communication to CSSRI, Karnal.

Sen, H.S., Ambast, S.K., Biswas, C.R., Mandal, A.B., Bandyopadhyay, B.K., Bal, A.R., Dutt, S.K., Maji, B. and Das, M. 1996. Hand Book on Improved Farming in Coastal Saline Soils of Sundarbans. Central Soil Salinity Research Institute, RRS, Canning. 16 p.

Sen, H.S., Bandyopadhyay, B.K., Maji, B., Bal, A.R. and Yadav, J.S.P. 2000. Management of coastal agro-ecosystem. In: Natural Resource Management for Agricultural Production in India (Yadav and Singh (Eds.)). New Delhi: 925-1022.

Septiningsih, E.M., Pamplona, A.M. Sanchez, D.L., Neeraja, C.N., Vergara, G.V., Heuer, S., Abdelbagi M. I. and Mackill, D.J. 2009. Development of submergence-tolerant rice cultivars: the Sub1 locus and beyond. Ann Bot.103: 151–160.

Sharma, D.K., Kumar, Ashok and Singh, K.N. 1990. Effect of irrigation scheduling on growth yield and evapotranspiration of wheat in sodic soils. Agricultural Water Management. 18: 267-276.

Sharma, D.K., Mishra, V.K. Singh, Y.P., Nayak, A.K., Singh, Ranbir, Kaledhonkar, M.J., Varshney, R.K. and Singh, Gurbachan 2008. Integrated Farming System Model for Waterlogged Sodic Soils. Technical Bulletin: CSSRI/ Lukhnow/01/2008, Central Soil Salinity Research Institute, Regional Research Station, Lucknow.12 p.

Sharma, D.K. and Singh, K.N. 1992. Effect of irrigation on growth, yield and evapotranspiration of mustard (*Brassica juncea*) in partially reclaimed sodic soils. Agricultural Water Management. 23: 225-232.

Sharma, D.K., Singh, K.N. and Chillar, R.K. 1985. Effect of time of first irrigation on yield of wheat in sodic soil. Indian Journal Agronomy, 30 : 459-463.

Sharma, D.P. and Gupta, S.K. 2006. Subsurface drainage for reversing degradation of waterlogged saline lands. Land Degradation and Development. (Accepted, On-line version available)

Sharma D.P., Singh K.N. and Rao K.V.G.K. 1992. Reuse of Saline Drainage Water for Irrigation. Better Farming Series No. 16. CSSRI, Karnal. 16 p.

Sharma, O.P. and Gupta, R.K. 1986. Comparative performance of gypsum and pyrites in sodic Vertisols. Indian J. Agric. Sci. 56:423-429

Sharma, S.K. and Bhargava, G.P. 1982. Ground water behavior and soil salinity in Haryana. Annual report. CSSRI, Karnal.

Singh, N.T. 1992. Salt affected soils in India. In: Land and Soils (Khoshoo, T.N. and Deekshatulu, B.L. Eds.). Indian National Science Academy, New Delhi. 65-100.

Shiva, V. 2003. Freedom from the Margins: Coastal Communities Celebrate the Golden Jubilee of India's Independence as Freedom from the Shrimp Industry http://www.navdanya.org/articles/acquaculture.htm

Singh, A. K. 2013. Building Climate Resilient Agriculture Production System. Lecture delivered at CSSRI on 44th Foundation day. CSSRI, Karnal. 22 p.

Singh, D. and Nijhawan, S.D. 1932. A study of the physico- chemical changes accompanying the process of reclamation in alkali soils. Indian J. Agric. Sci. 2:1-18.

Singh, G. and Dagar, J.C. 1998. Agroforestry in salt affected soils. In: Agricultural Salinity Management in India (Tyagi and Minhas Eds.). CSSRI, Karnal. 473-487.

Singh, G., Dagar, J.C. and Singh, N.T. 1997. Growing fruit trees in a highly alkali soils – a case study. Land Degradation and Development. 8: 257-268.

Singh, G. and Dagar, J.C. 2005. Greening Sodic Soils: Bichhian Model. Technical Bulletin No.2 /2005. Cental Soil Salinity Research Institute, Karnal. 51p.

Singh, G., Singh, N.T. and Tomar, O.S. 1993. Agroforestry in Salt-affected Soils. Technical bulletin No. 17, CSSRI, Karnal. 65 p.

Singh, I.S., Singh. R.K. and Khanna, S.S. 1994. Management of Sodic Soils through Plantations. Bulletin, N.D. University of Agriculture and Technology, Faizabad, UP. 51 p.

Singh K.N., Sharma D.P. and Rao, K.V.G.K. 1992. Technology for Crop Production in Waterlogged Saline Soils. Better farming in Salt Affected Soils, Ext. Series. 17: 12 p.

Singh K.N., Swarup A, Sharma D.P. and Rao K.V.G.K. 1990. Influence of subsurface drainage and phosphorus application on soil salinity and uptake of phosphorus by wheat in saline soil. J. Indian Soc. Soil Sci. 38: 511-515.

Singh K.N., Swarup A., Sharma D.P. and Rao K.V.G.K. 1992. Effect of drain spacing and phosphorus levels on yield, chemical composition and uptake of nutrient by Indian mustard (*Brassica juncea*). Experimental Agriculture. 28: 135-142.

Singh, M.V. and Abrol, I.P. 1986. Split application of zinc better than a single dose. Indian Farming. 36: 35-37.

Singh, N.T. 1974. Physico-chemical changes in sodic soils incubated at saturation. Pl. Soil. 40: 303-311.

Singh, N.T. 1998. Reclamation and management of alkali soils. In: Agricultural Salinity Management in India (Tyagi and Minhas (Eds.)). CSSRI, Karnal. 261-278.

Singh, N.T., Hira, G.S. and Bajwa, M.S. 1981. Use of amendments in reclamation of alkali soils in India. Agro. Es. Talaj. 30:158-177.

Singh, P. 1980. An appraisal of the economics of reclamation of alkali soils. Proc. Int. Symp. Salt Affected Soils, Karnal. 555-561.

Singh O.P. 1993. Drainage problems and design criteria for land drainage symptoms. Proceeding of the National Workshop on "Sustainable Irrigation in Saline Environment" Feb. 17- 19, CSSRI, Karnal. 204-218.

Singh, Y.P., Sharma, D.K., Singh, Gurbachan, Nayak, A.K., Mishra, V.K. and Singh R. 2008. Alternate Land Use Management for Sodic Soils. Technical Bulletin 2/2008. CSSRI Regional Research Station, Lucknow. 16 p.

Singhla, S.K., Bhargava, G.P. and Gupta, I.C. 1974. Relationships between certain diagnostic indices of salt affected soils of Indo-Gangetic alluvial region. Ann. Arid Zone 13: 218-223.

Smedema, L.K. and Rycroft, D.W. 1983. Land Drainage: Planning and Design of Agricultural Drainage Systems. Cornell University Press, Ithaca, New York. 375 p.

Smith, R.J. and Hancock, N.H. 1986. Leaching requirement of irrigated soils. Agricultural Water Management. 11:13-22.

Sondhi, S.K. 2007. Subsurface drainage: a case study of southwest Punjab. In: Agricultural Land Drainage: Reclamation of Waterlogged Saline Lands (Ambast, Gupta and Singh, Eds.). Central Soil Salinity Research Institute, Karnal and CAD&WM, New Delhi. 116-127.

Suarez, D.L. 1975. Precipitation of magnesium carbonates. Annual Report U.S. Salinity Laboratory, Riverside, Ca, USA.

Suarez, D.L. 1981. Relationship between pH_c and SAR and an alternate method for estimating SAR of soil or drainage water. Soil Sci. Soc. Am. J. 45:469-475.

Subbaiah, G.V., Gupta R.K., Rao K.V.G.K. and Subba Rao I.V. 1991. Development of doruvu technology for skimming fresh water floating over saline ground water in coastal sands of Andhra Pradesh J. Indian Soc. Coastal agric. Res. 9: 205-206.

Subba Rao, P.V. and Mantri, V.A. 2006. Indian seaweed resources and sustainable utilization: scenario at the dawn of a new Century. Current Science. 91: 164-174.

Sundara, B. and Vasantha, S. 2004. Sugarcane Management in Saline Soils. Ext. Pub. 81. Sugarcane Breeding Institute, Coimbatore.

Swarup, A. and Yaduvanshi, N.P.S. 2004. Response of rice and wheat to organic and inorganic fertilizers and soil amendment under sodic water irrigated conditions. IRRI Newsletter. 29.1: 49-51.

Talati, R.P. 1941. Damaged lands in the Deccan and their classification. Indian J. Agric. Sci. 11: 959- 977.

Talati, R.P. 1947. Field experiments on the reclamation of salt lands in Baramati of Bombay Deccan. Ind. J. Agric. Sci. 17: 153- 174.

Talsma, T. 1966. The effect of soil physical condition on reclamation of saline land. 6th Cong. Irrig. and drainage. New Delhi 19:83- 89, 91.

Talsma, T. 1963. The Control of Saline Groundwater. Meded Landb Hoogesch, Wageningen. 63:1-68.

Tejawat, C.M. 2007. Reclamation of waterlogged saline soils through drainage in Chambal command area, Rajasthan. In: Agricultural Land Drainage-Reclamation of Waterlogged Saline Lands (Eds. Ambast, Gupta and Singh). Command Area Development and Water Management, New Delhi and CSSRI, Karnal. 128-142.

Thokal, R.T. 1990. Strategies of leaching in highly saline soils. Unpublished M. Tech. Thesis, G.B.P.U.A.&T., Pantnagar.

Tiwari, K.N., Sharma, D.N. and Tripathi, S.K. 1989. Salt-Affected Soils of U.P.: Their Reclamation and Management. C. S. Azad University of Agriculture and Technology, Kanpur.

Tripathi, R.S. and Sharma, D.K. 2011. *Kshariya Bhumi Sudhar evam uske samajik-aarthik Labh.* Tech. Bull. CSSRI/Karnal/2009/02. Central Soil salinity Research Institute, Karnal. 12 p.

Tomar, O.S. and Gupta, R.K. 1984. Annual Report, CSSRI, Karnal. 82-86.

Tomar, O.S. and Gupta, R.K. 1985. Performance of some tree species in saline soils under shallow and saline water table conditions. Plant and Soil. 27: 329-335.

Tomar, O.S. and Gupta, R.K. 1987. Annual Report, CSSRI, Karnal. 101-103.

Tomar, O.S. and Gupta, R.K. 1992. Performance of some exotic tree species in saline soils. Annual Report, Central Soil Salinity Research institute, Karnal. 98-99.

Tomar, O.S., Minhas, P.S., Sharma, V.K. and Gupta, R.K. 2003. Response of nine forage grasses to saline irrigation and its schedules in a semi-arid climate of north-west India. Journal of Arid Environment. 55: 533-544.

Tomar, O.S., Minhas, P.S., Sharma, V.K., Singh, Y.P. and Gupta, Raj K. 2002. Performance of 32 tree species and soil conditions in a plantation established with saline irrigation. Forest Ecology & Management. 177: 333-346.

Tomar, O.S. and Yadav, J.S.P. 1980. Effect of saline irrigation water of high salinity with and without application of fertilizers on the growth of *Eucalyptus hybrid.* Indian J. Range Mgmt. 2:121-127.

Tomar, O.S. and Yadav, J.S.P. 1985. Effect of saline water irrigation and fertilizer application on growth of tree seedlings. Annals of Arid Zone. 24: 94-100.

Tyagi, N.K. 1976. Studies on performance of closed level and graded borders in alkali soils. Ann. Rep. CSSRI, Karnal.

Tyagi, N.K., Gupta, S.K. and Datta, K.K. 1996. Rural Land Development through Alkali Land Reclamation: The Gurusikaran Model. CSSRI, Karnal and AFPRO, New Delhi. 33 p.

Tyagi, N.K. and Singh, O.P. 1979. Investigations on effect of soil moisture on performance of land levelling implements. J. Agric. Engg. 15: 67-72.

US Bureau of Reclamation, SCS. 1971. Soil Conservation Service. National Engineering Handbook, Section 16: Drainage of Agricultural Lands, USDA, Washington, D.C.

US Salinity Laboratory Staff (USSLS). 1954. Diagnosis and Improvement of Saline and Alkali Soils. Handbook 60. USDA, Washington DC. 160 p.

van Der Molen, W.H. 1956. Desalinization of saline soils as a column process. Soil Sci. 81:19-27.

van Der Molen, W.H. 1973. Salt balance and leaching requirement. In: Drainage Principles and Applications: Theories of Field Drainage and Watershed Runoff. Int. Institute for Land Reclamation and Improvement. Wageningen. 11:62- 99.

van Hooren, J.W. 1979. The effect of capillary flow on salinization and the concept of critical depth for determining drain depth. In: J. Wesseling (Ed.). Proc. of the Int. Drainage Workshop. ILRI Pub. 25, Wageningen. 686-700.

Verma, S.K. and Gupta, R.K. 1989. Leaching behavior of saline clay soil under two modes of water application. J. Indian Soc. Soil Sci. 37:803-808.

Working Group. 1991. Report of the Working Group on Problem Identification in Irrigated Areas with Suggested Remedial Measures. Ministry of Water Resources, Government of India, New Delhi.

Yadav, J.S.P. and Agarwal, R.R. 1959. Dynamics of soil changes in the reclamation of saline alkali soils of the Indo-Gangetic alluvium. J. Indian Soc. Soil Sci. 7:213-222.

Yadav, J.S.P., Bandyopadhyay, A.K. and Bandyopadhyay, B.K. 1983. Extent of coastal saline soils of India. Journal of Indian Society of Coastal Agricultural Research. 1: 1-6.

Yadav J.S.P., Bandyopadhya, A.K., Rao, K.V.G.K., Sinha, T.S. and Biswas, C.R. 1979. Coastal Saline Soils of India. Bulletin No. 5. CSSRI, Karnal. 34p.

Yadav, J.S.P. and Girdhar, I.K. 1977. Salt translocation and crop growth under sprinkler and surface methods of irrigation in sodic soils. Indian J. Agric. Sci. 47:397-400.

ANNEXURE I

IRRIGATION WATER SALINITY LIMITS FOR SELECTED CROPS

Crop/Location	Soil type	No. of Years	Previous crop	EC_{iw} (dS/m) for relative yield		
				90 %	75 %	50 %
CEREALS						
Wheat						
Agra	Sl	6	Pearl millet	6.6	10.4	16.8
Agra (late sown)	Sl	2	Toria	4.3	6.6	11.0
Dharwad	Scl	5	Sorghum	3.4	7.0	12.9
Hisar	Sl	5	Sorghum/fallow	6.1	8.7	13.0
Indore	Cl	8	Maize	4.7	8.7	15.2
Jobner	Ls	4	Fallow	8.3	11.7	17.5
Jodhpur	Scl	2	Fallow	8.3	10.2	13.4
Karnal	Sl	5	Fallow	9.1	10.8	13.7
"	S	5	Fallow	14.0	16.1	19.5
Barley						
Agra	Sl	2	Fallow	7.2	11.3	18.0
Jobner	Ls	5	Fallow	13.0	21.1	-
Rice						
Agra	Sl	1	*Berseem*	2.3	4.6	8.6
Bapatla	Scl	6	*Kharif* rice	2.2	3.9	6.8
		3	*Rabi* rice	1.8	2.9	4.8
Maize						
Dharwad	Scl	5	Sorghum	3.7	7.8	14.5
Indore	Cl	7	Wheat	2.2	4.7	8.8
Pearl millet						
Jobner	Ls	5	Fenugreek	11.9	22.7	-
Agra	Sl	4	Wheat	5.4	9.0	15.0

Italian-millet						
Bapatla	S	5	Sunflower (K)	2.4	4.6	8.2
	S	4	Italian-millet (R)	2.5	4.9	8.7
Sorghum						
Dharwad	Scl	6	Wheat	2.6	5.1	9.2
Agra	Sl	3	Mustard	7.0	11.2	18.1
Sorghum (fodder)						
Agra	Sl	2	*Berseem*	5.2	10.2	18.4
Hisar	Sl	2	Wheat	2.5	5.7	11.0
OIL SEEDS						
Mustard						
Agra	Sl	6	Sorghum	6.6	8.8	12.3
Bapatla	Scl	5	Soybean	3.8	7.9	14.7
Jobner	Ls	2	Cluster bean	6.6	13.5	-
Toria						
Agra	Sl	1	Wheat	4.7	5.1	5.9
Safflower						
Dharwad	Scl	5	Maize	3.3	6.8	12.6
Sunflower						
Bapatla	Sl	3	Mustard	3.5	7.2	13.4
Hisar	Sl	1	Wheat	2.5	5.7	11.0
Groundnut						
Bapatla	S	5	Italian millet	1.8	3.1	5.3
Soybean						
Bapatla	Scl	3	Mustard	2.0	3.1	5.0
PULSES/LEGUMES						
Pigeon pea						
Agra	Sl	6	Onion	1.3	2.3	3.9
Cluster bean						
Bapatla	Sl	3	Variable	3.2	4.5	6.8
Jobner	Ls	2	Mustard	3.9	6.6	11.1
Cowpea						
Jobner	Ls	2	Variable	8.1	13.2	-
Berseem						
Agra	Sl	5	Rice/sorghum	2.5	3.2	4.4

VEGETABLES						
Onion						
Agra	Sl	5	Variable	5.1	6.0	7.5
Bapatla	S	5	Pigeon pea	1.8	2.3	3.3
Potato						
Agra	Sl	5	Okra	2.1	4.3	7.8
Tomato						
Bapatla	S	3	Variable	2.4	4.1	6.9
Okra						
Agra	Sl	2	Variable	2.1	3.9	6.7
Bapatla	S	5	Potato	2.7	5.6	10.5
Brinjal						
Bapatla	S	2	Variable	2.3	4.1	7.1
Fenugreek						
Jobner	Ls	3	Pearl millet	3.1	4.8	7.6
Chillies						
Bapatla	S	2	Variable	1.8	2.9	4.9
Jobner	Ls	2	"	4.5	7.5	12.5
Coriander						
Bapatla	S	3	Variable	2.9	5.8	10.7
Jobner	Ls	2	"	9.8	15.4	-
Bittergourd						
Bapatla	S	3	Variable	2.0	3.4	5.8
Bottlegourd						
Bapatla	S	3	Variable	3.2	4.5	6.8

S, sandy; ls, loamy sand; sl, sandy loam; l, loam; scl, sandy clay loam; cl, clay loam

The annual rainfall in Agra, Bapatla, Dharwad, Indore, Jobner, Hisar, and Karnal was 660, 803, 778, 750, 500, 450 and 850 mm, respectively.

Source: Minhas and Gupta (1992)

WATER ANALYSIS TO CALCULTE ADJ. R_{Na}

Water analysis of a given water sample is

Ca	=	2.54 meq/L	Mg	=	1.34 meq/L
Na	=	6.87 meq/L	EC	=	1.0 dS/m
CO_3	=	0.38 meq/L	HCO_3	=	4.21 meq/L

Find RSC, SAR and adj. R_{Na}

$$RSC = (CO_3 + HCO_3) - (Ca + Mg) \text{ meq/L}$$

$$=(0.38+4.21) - (2.54+1.34) = 4.59 - 3.88 = 0.71 \text{ meq/L}$$

$$SAR = \frac{Na}{\sqrt{(Ca + Mg)/2}}$$ Where Na, Ca and Mg are in meq/L

$$= \frac{6.87}{\sqrt{(2.54 + 1.34)/2}} = 4.93$$

$$Adj.R_{Na} = \frac{Na}{\sqrt{(Ca_x + Mg)/2}}$$

EC_{iw} is 1.0 dS/m, HCO_3/Ca ratio 4.21/2.54 = 1.66

Ca_x value from Table A-1 corresponding to the nearest values of HCO_3/Ca and EC_{iw} is 1.38

$$= \frac{6.87}{\sqrt{(1.38 + 1.34)/2}} = 6.87 / 1.18 = 5.87$$

Table A-1. Calcium concentration (Ca_x) expected to remain in near-surface soil-water following irrigation (meq/L)

Ratio of HCO_3/Ca	Salinity of applied water (EC_{iw}) (dS/m)											
	0.1	0.2	0.3	0.5	0.7	1.0	1.5	2.0	3.0	4.0	6.0	8.0
0.05	13.20	13.61	13.92	14.40	14.79	15.26	15.91	16.43	17.28	17.97	19.07	19.94
0.10	8.31	8.57	8.77	9.07	9.31	9.62	10.02	10.35	10.89	11.32	12.01	12.56
0.15	6.34	6.54	6.69	6.92	7.11	7.34	7.65	7.90	8.31	8.64	9.17	9.58
0.20	5.24	5.40	5.52	5.71	5.87	6.06	6.31	6.52	6.86	7.13	7.57	7.91
0.25	4.51	4.65	4.76	4.92	5.06	5.22	5.44	5.62	5.91	6.15	6.52	6.82
0.30	4.00	4.12	4.21	4.36	4.48	4.62	4.82	4.98	5.24	5.44	5.77	6.04
0.35	3.61	3.72	3.80	3.94	4.04	4.17	4.35	4.49	4.72	4.91	5.21	5.45
0.40	3.30	3.40	3.48	3.60	3.70	3.82	3.98	4.11	4.32	4.49	4.77	4.98
0.45	3.05	3.14	3.22	3.33	3.42	3.53	3.68	3.80	4.00	4.15	4.41	4.61
0.50	2.84	2.93	3.00	3.10	3.19	3.29	3.43	3.54	3.72	3.87	4.11	4.30
0.75	2.17	2.24	2.29	2.37	2.43	2.51	2.62	2.70	2.84	2.95	3.14	3.28
1.00	1.79	1.85	1.89	1.96	2.01	2.09	2.16	2.23	2.35	2.44	2.59	2.71
1.25	1.54	1.59	1.63	1.68	1.73	1.78	1.86	1.92	2.02	2.10	2.23	2.33
1.50	1.37	1.41	1.44	1.49	1.53	1.58	1.65	1.70	1.79	1.86	1.97	2.07
1.75	1.23	1.27	1.30	1.35	1.38	1.43	1.49	1.54	1.62	1.68	1.78	1.86
2.00	1.13	1.16	1.19	1.23	1.26	1.31	1.36	1.40	1.48	1.54	1.63	1.70
2.25	1.04	1.08	1.10	1.14	1.17	1.21	1.26	1.30	1.37	1.42	1.51	1.58
2.50	0.97	1.00	1.02	1.06	1.09	1.12	1.17	1.21	1.27	1.32	1.40	1.47
3.00	0.85	0.89	0.91	0.94	0.96	1.00	1.04	1.07	1.13	1.17	1.24	1.30
3.50	0.78	0.80	0.82	0.85	0.87	0.90	0.94	0.97	1.02	1.06	1.12	1.17
4.00	0.71	0.73	0.75	0.78	0.80	0.82	0.86	0.88	0.93	0.97	1.03	1.07
4.50	0.66	0.68	0.69	0.72	0.74	0.76	0.79	0.82	0.86	0.90	0.95	0.99
5.00	0.61	0.63	0.65	0.67	0.69	0.71	0.74	0.76	0.80	0.83	0.88	0.93
7.00	0.49	0.50	0.52	0.53	0.55	0.57	0.59	0.61	0.64	0.67	0.71	0.74
10.00	0.39	0.40	0.41	0.42	0.43	0.45	0.47	0.48	0.51	0.53	0.56	0.58
20.00	0.24	0.25	0.26	0.26	0.27	0.28	0.29	0.30	0.32	0.33	0.35	0.37
30.00	0.18	0.19	0.20	0.20	0.21	0.21	0.22	0.23	0.24	0.25	0.27	0.28

Adapted from Suarez (1981); Assumes a soil source of calcium from lime ($CaCO_3$) or silicates; no precipitation of magnesium, and partial pressure of CO_2 near the soil surface (pCO_3) is 0.0007 atmosphere; Ca_x, is in meq/L

SUBJECT INDEX